Designing intranets

Creating sites that work

James Robertson

Designing intranets: creating sites that work
by James Robertson

Published by Step Two Designs
PO Box 551
Broadway NSW 2007
Australia

www.steptwo.com.au
contact@steptwo.com.au

ISBN: 978-0-9808131-0-4

Table of contents

Chapter 1

Introduction

Websites are easy to design. There is a whole world of other websites to see, access and evaluate. Best practices are well understood, and widely written about in books and blogs. Innovation is driven by movements such as web 2.0, producing a never-ending stream of new sites and ideas that feed the rest of the web ecosystem.

Not sure what your website should look like? Have a look at other similar sites, or the 'leading' sites of the day. How does your site compare? Pick a cross-section of peers or competitors, and evaluate the sites based on whatever criteria makes the most sense.

There's even a rule coined by the usability guru Jakob Nielsen which states "Users spend most of their time on other sites. This means that users prefer your site to work the same way as all the other sites they already know."

It's clear when a website falls behind, and simple to demonstrate to management that change is required. When it comes to the site redesign, there is no shortage of professionals with extensive web experience.

Now this grossly underplays the challenges in delivering a great website, and borders on disrespectful to the web professionals who bring so much experience to bear. It does, however, start to highlight the many differences between designing websites and intranets.

Intranets are no less important than public-facing websites. They are relied on every day by staff as they conduct their work. It can easily be argued that organisations will struggle to be successful without the behind-the-scenes support of a great intranet. Despite this, designing intranets still borders on being a black art, and is far less understood than website design.

Figure 1-1: An intranet homepage inspired by newspaper websites and modern design approaches. Screenshot courtesy of AMP.

Figure 1-2: A very different approach to intranet homepage design, focusing on personalisation and task completion. Screenshot courtesy of Framestore.

The immediate challenge confronting anyone designing (or redesigning) an intranet is seeing other sites. Hidden away within organisations, intranets by definition are not accessible to the general public. This makes it very hard to identify best practices or to discern patterns.

This an exciting time for intranets, with new approaches being explored across the globe. Figures 1-1 and 1-2, for example, show very different ways of designing intranet homepages.

Despite this, there is considerable 'reinventing the wheel', with the same mistakes made and lessons learnt. This makes intranets uncharted territory for many teams, despite the decade or more that intranets have been in place.

As the pace of business increases, there is constant pressure to 'do more with less'. Staff need to work efficiently, and business practices need to be streamlined and repeatable.

Intranets play a key role in meeting these needs, with the importance of knowledge and collaboration being taken for granted in most organisations. There is, however, a big gap to close.

It's safe to say that many staff don't love their intranets. Quite the opposite, with intranet teams often flooded with complaints that 'I can't find anything'. Senior management recognition and intranet resources are also poor.

No organisation would consider the possibility of turning the intranet off, but it can sometimes feel like an unloved child. The starting point to resolving this is to deliver an intranet that actually works well for staff. Information can be quickly found when needed, tools can be easily understood and used.

Good intranet design underpins site success. More than the colours and logos, this involves site structure (how staff navigate) and page layout (how they work out what to click on). If these fundamentals aren't right, no amount of new technology or functionality will help.

This book aims to shed light onto the practice of intranet design. It will uncover and share the fundamental principles, methodologies and approaches that underpin successful intranet designs. It will reduce much of the confusion and uncertainty that surrounds intranet projects.

The overall goal is to get the 'basics' out of the way, to bring all intranets up to an acceptable level of design and functionality. This frees up intranet teams, and the industry as a whole, to move onto the next round of more exciting improvements.

Intranets are still at the early stages of their journey, and the best is yet to come. Intranets of the future will play a pivotal role in supporting day-to-day work for all staff, but the starting point must be to create sites that work at a fundamental level.

My promise to you

Everyone has an opinion about intranets. Staff 'know what they like', stakeholders 'know what they want', and every expert has simple rules to follow when designing sites.

The aim in writing this book is to bring greater clarity to intranet design, not to add yet another set of opinions to the sea of debate. When presenting material, I therefore make the following promises:

- if there is a clear principle or rule to follow, I'll state it (thereby providing 'the answer')

- otherwise, I'll provide a clear methodology to follow or a set of techniques to use ('how to work out the answer')

- in some cases, there may be more than one solution or approach, in which case we'll explore them ('here are the options')

- in all cases, I'll avoid providing simple rules where none apply, or a single answer where the options are less clear

That is not to say that I don't have opinions on what works, and what doesn't (ask anyone who has sat through one of my workshops!). I will endeavour, however, to be clear that what I'm stating is simply that: an opinion.

Too many simple rules lead to undesirable outcomes, even when the original intentions are good. Too many principles prevent good teams thinking through the options, or considering different approaches.

Throughout the book, I will always provide pragmatic advice, based on what works in the real-life complexities of modern organisations. Step Two Designs, the consulting firm that I lead, has worked with many different organisations across every sector.

We've seen many hundreds of intranets the world over, at conferences, workshops and events. We run the global Intranet Innovation Awards, uncovering great intranet ideas. We've talked with intranet managers and teams about their challenges, and have heard many stories of success and failure.

Intranet design does not happen in isolation, independent of overall strategy, internal politics, organisational culture, technology capabilities, or staff preferences. In many cases, these practical considerations determine the 'best' intranet design, independent of design principles.

I'll therefore return to these to provide guidance when the way is unclear, or when teams have many potential avenues to explore.

Robertson

Audience for this book

This book is written for anyone who has been given the task of designing a new intranet from scratch, or redesigning an existing intranet. This might include:

- centralised intranet teams

- business unit owners of intranet sites or sections

- web or web services teams

- communications teams

- IT teams

- project teams assigned to an intranet redesign

- an individual given the intranet redesign responsibility

Note that these are all in-house groups (or individuals) consisting primarily of business people, communications or IT professionals. No assumptions have been made about prior intranet experience, and this book pays particular attention to covering the fundamentals and key concepts.

The book is all about giving teams the tools, skills and knowledge they need to succeed. There is no single group that does (or should) own the intranet, and intranet teams come from varied backgrounds.

Regardless of the different skill sets this implies, this book should give the information required to make an intranet design or redesign project less painful and more successful.

Beyond these audiences, there are a range of professionals with greater intranet experience, such as:

- usability, information architecture and user experience professionals

- consultants and consultancies

- dedicated intranet professionals

This book is not written for these audiences. Most of these groups will already have a strong grasp of core methodologies, as well as having a broad exposure to intranets and intranet projects.

These audiences may still find value in this book, and are encouraged to skip the fundamentals and move straight to the intranet-specific discussions. The screenshots and examples will be particularly valuable.

Fundamentals and beyond

There are those in the industry who argue that intranets are dull and narrow-minded. Mere dumping ground for corporate documents, intranets have a decreasing relevance in today's modern organisations.

These criticisms have some merit. When intranets consist solely of thousands of pages of content, intranet teams struggle to find the right structure and navigation. With such a narrow role in the organisation, intranets may not generate the levels of use that are desired.

The rise of web 2.0, enterprise 2.0 and social media have all highlighted the long-overlooked role of people and community. They have also kick-started a new era of innovation, much of which has yet to be seen on intranets.

Despite these concerns, this book retains its focus on the fundamentals of intranet design, on how to make the intranets we have today work well. This is a carefully considered decision.

Having observed countless intranets and intranet teams, it seems there are several hurdles that must be overcome for these sites to succeed:

- Basic usability and design principles need to be applied to all sites, to ensure that intranets work well for staff.

- Knowledge of what works (and what doesn't) needs to be shared more widely amongst the industry.

- Common problems need to be resolved in order to make time for intranet innovation.

- Experienced designers need to be attracted to intranets, repositioning these sites as something worthy of the best thinking.

This book aims to be a 'how-to guide' for members of intranet project teams, spelling out the fundamentals of a methodology that should work in any organisation.

Don't stop there, however. What is outlined in this book is just the beginning, and there is considerable new territory yet to explore. My hope is that by addressing the immediate concerns and needs of staff, intranet teams will finally have enough time and mind-share to truly innovate.

To start this process, a number of more advanced topics will be covered in Chapter 19. These highlight where the edges of the 'known' lie, and therefore where the exploration should begin.

Structure of the book

This book is written as a how-to guide on designing intranets, focusing on techniques and methodologies wherever possible. Starting from basic principles, you will be taken from the start of the design process right through to the end:

- *Chapters 2 and 3* set the scene, outlining the importance of good intranet design, and highlighting common mistakes made when designing sites.

- *Chapters 4, 5 and 6* focus on defining the purpose, scope and strategy of the intranet. This starts with an understanding of staff needs. By the end of this work, the intranet team has a clear plan of attack for the project.

- *Chapters 7 to 16* are the heart of the book, providing a step-by-step guide to a practical methodology for designing an intranet that works well for staff, including key techniques such as card sorting and usability testing.

- The remaining chapters cover off a number of practical considerations when designing intranets, as well as touching upon a number of advanced and future topics to be addressed by intranet teams.

By the end of the book, you should have a clear idea of how to design the intranet with confidence. Armed with key hands-on techniques, draft designs can be validated and refined, allowing you to escape the often conflicting opinions of the intranet's many stakeholders.

Technology agnostic

Intranets can be published and managed using many different types of technologies, including web content management systems, portals, enterprise platforms such as SharePoint or collaborative tools. There are literally thousands of products worldwide, many of which are well-suited for intranet uses. Intranets can even by published 'by hand', directly editing HTML pages.

While this book is founded on a deep understanding of technology products and options, it is designed to be relevant regardless of the tools used. While each technology will have a different philosophy and architecture, the fundamental principles outlined in the pages to come can be easily adapted to fit.

Where recent technologies are providing new options for intranets, these will be flagged at the relevant point. The screenshots are also drawn from organisations using a wide mix of tools, helping to give a broad cross-section of examples for intranet projects.

The intranet team

Throughout this book, I will be referring to the 'intranet team', or sometimes the 'project team'. Who are they?

It is safe to say that intranets don't run themselves. They require someone to manage and nurture them, and to grow and enhance them over time. Without this, intranets invariably become unstructured, out of date, unused and useless.

This team (or individual) may sit in many possible locations in the organisation, with communications, HR, knowledge management and IT all being common homes. There may be an 'intranet manager', an 'internal communications specialist', 'web channel manager', 'knowledge manager', 'online communications lead' or one of a hundred other titles.

My statement on this is as follows: "I don't care where the intranet team sits in the organisation as long as they have the right skills and focus, and the support of their management".

In a small organisation, running the intranet might be a part-time responsibility of a single individual. Once an organisation grows beyond a few hundred staff, there will be a need for a full-time role dedicated to the intranet. Over a thousand staff, the intranet will require a permanent team. Very large organisations can have upwards of a dozen staff devoted solely to running the intranet.

On a practical note, the approach outlined in this book will require work, a lot of it. Squeezing this in as a part-time (added) responsibility will be difficult. Busy intranet teams may even have to bring in additional project members for the duration of a redesign project.

The work doesn't end when the design or redesign project is complete. Intranets can easily slide back into disrepair, undoing all of the improvements delivered by the project. A one-off redesign should also be the start of an ongoing journey of improvement and enhancement for the intranet, which is often where the greatest value is added.

For all these reasons, I will assume that there is a long-term team in place to design and manage the intranet. For the sake of simplicity, I will call this group the 'intranet team'. throughout the book.

If there isn't a formal intranet team, don't despair! Intranet projects can be scaled to fit the available resources, even if this is just 25 per cent of the time of a single individual. Incremental approaches can also be very effective in these situations, using the techniques outlined in this book, but applying them section-by-section to the site. Hopefully, delivering a great new intranet that works well for staff, and is valuable for the organisation as a whole, will help to justify further resources in the future.

For more on the roles and responsibilities of intranet teams, see the many articles published by Step Two on this topic:

www.steptwo.com.au/tag/intranet-teams

A note on screenshots

As discussed earlier, one of the biggest challenges for intranet teams is finding ways to see other intranets. For this reason, many screenshots have been included throughout this book, to illustrate common approaches, but also just to give teams an opportunity to see what other sites look like.

Many intranet teams from around the world have been very generous in sharing these screenshots. Like every site, none of these intranets is perfect. Every one does, however, reflect some great ideas and approaches, and this is what is being highlighted.

Learn from these screenshots, but don't copy them directly. Take individual elements, and find what will work in your organisation. Most of all, thank the brave souls who have shared their work for all to see.

Acknowledgements

My role is to work with intranet teams, to uncover best (and worst) practices, and to share these with the wider community. At Step Two Designs, I'm lucky enough to have a number of true experts in usability and information architecture on my team. Over the years, we've also had some of the best in the industry working with us.

I'd therefore like to acknowledge the considerable contribution to our body of knowledge made by everyone who has been a part of the Step Two team. In particular, special thanks must go to Tina Calabria, Iain Barker, Patrick Kennedy, Donna Spencer, and Rebecca Rodgers.

Thanks also to those who generously read various versions of this book, and provided invaluable feedback. This includes Gerry Gaffney, Lukas Karrer, Patrick Kennedy, Sam Marshall, Harry Max, Jane McConnell, Maish Nichani, Nic Price, Martin White and Andrew Wright.

A brief note on spelling

Readers in the US may be initially confused by the spelling used throughout this book. I'm Australian, and we use Australian English in our writing. For example, this means 's' instead 'z', 'colour' instead of 'color', 'centre' instead of 'center'.

I am confident that the meaning of the book will shine through clearly. There is no one way of writing that would be universally seen as 'right' the world over, so the material in this book will be presented in my native tongue and spelling.

Chapter 2

Intranet design matters

Talking to staff in most organisations, the most common complaint about the intranet is: "I can't find what I want". Regardless of whether they are searching or browsing, staff consistently struggle to locate what they need, when they need it.

This is not a new challenge. Intranets are now well over a decade old, and many organisations have redesigned their intranets several times. These problems are not due to a lack of intent or effort; intranet teams have consistently been working hard on delivering an intranet that works well. And yet the complaints remain.

There are many reasons why staff struggle with intranets, commonly arising from design mistakes that will be covered in Chapter 3. There are also practical methodologies that can be used to deliver intranets that work.

Does it matter? Do the complaints of staff really need to be addressed?

There are few organisations that don't have an intranet, and none would seriously entertain the notion of turning the intranet off. Quietly, without fanfare, the intranet has become the definitive source of corporate information. From HR to finance and IT, the intranet is the home for policies, procedures and forms.

Beyond this core corporate content, intranets grow to encompass key business unit information and front-line content. This is the information that staff need daily to do their work.

For example, a 2007 study from Accenture showed that managers spend up to two hours a day searching for information, and more than 50 per cent of the information they obtain has no value to them. That's a lot of wasted time!

More fundamentally, why spend all the time and money on maintaining an intranet if it doesn't work well for staff? There is little point in publishing thousands of pages of content, if they are never found or used.

Staff complaints are legitimate. While some intranet teams and content owners would like to believe otherwise, staff are not complaining without just cause. We have seen many hundreds (perhaps thousands) of intranets, and most are plainly hard to use.

The intranet also competes with every other way of finding information, and not just in electronic form. As a staff member, I'll ask the person next to me first, and then query Peter in the cubicle across the corridor. I'll pick up the phone to ring Jane, who's been here for years, and will double-check that I didn't file away a useful message in my email program.

If all these efforts fail, I'll go to the intranet. Fundamentally, staff can (and should) use the quickest and easiest method to find the information they need. To force them to do otherwise would be to institutionalise inefficient practices.

This puts a lot of pressure on intranets. They must be more than just a collecting place for content. They must be quick, easy, effective and satisfying to use. If they don't measure up to these criteria, perhaps we'd be better off without our intranets.

Going beyond words and documents, intranets must also help staff to complete tasks. They should streamline common business processes, saving time, reducing mistakes and eliminating manual handling. They must become a 'place for doing things', not just for 'reading things'.

The good news is that we *can* deliver intranets that work well. That so many intranets currently fail is a reflection of the slow spread of knowledge between organisations, and the challenges of managing sites hidden away from public view.

Intranets are not impossibly hard to do right. They are not broken by design, and they can work well for staff if designed in the right way.

Best practices are now well understood, and there are practical techniques that can be used to ensure what is delivered works. Following a sound methodology allows teams to confidently produce designs that meet both staff and organisational needs. Design projects can be more productive, not to mention more enjoyable for the project team.

There will always be challenges, but we can at least get the basics right, giving us more time to focus on delivering the greatest business value with our intranets.

What is an intranet?

Before going further, it's probably useful to put some shape around what we mean we use the word 'intranet'. The following working definition may prove useful:

> *Intranets are the internal web-based environments that provide staff with the tools and information they need to do their job.*

In practice, it can be useful to look at this through the lens of five fundamental purposes:

- *Content.* The source of policies, procedures and other information staff need to make decisions and complete common tasks.

- *Communication.* A channel to deliver news and other messages to most (ideally all) staff throughout the organisation.

- *Collaboration.* A platform to connect people with people, and a space for them to work together online.

- *Activity.* A 'place for doing things' and not just for 'reading things', allowing common and important tasks to be completed online.

- *Culture.* A site that reflects the current culture of the organisation, as well as supporting and fostering changes towards a desired future culture.

If this still seems to be a very broad definition that could encompass many different internal systems and platforms, you'd be right. There isn't, and perhaps shouldn't be, a clear line between intranets and other enterprise technologies.

From a staff perspective, the intranet can often be 'the thing you access via the blue e' (the Internet Explorer icon on the desktop). Staff don't distinguish between systems, and don't want to.

From this point of view, everything could be seen as the intranet. While this is conceptually simple for staff, it introduces many challenges for the behind-the-scenes owners of content and tools. For example, if the intranet team owns the homepage, and HR owns the employee self-service tool, who then is responsible for 'the intranet'?

Intranets are most valuable when they find a balance between the five purposes outlined above. They are also most useful for staff when they make things simpler, blurring the lines between systems and silos.

While this is above all a question of intranet strategy, the overall purpose of the site should guide every step of the design process. Care should also be taken not to focus too much on just a few aspects of the site.

Intranets and websites

Intranets and public-facing websites are superficially similar, both publishing content as web pages, and managed using similar technologies. Beyond this, however, websites and intranets are different in almost every respect, as shown in Table 2-1.

The primary goals of corporate websites are often to communicate information, support marketing, build brand or support e-commerce. Website goals are outwards-focused, and aligned with marketing or public affairs strategies.

Intranets are designed to meet the needs of staff, as well as organisational strategic goals. Intranet goals are usually defined in terms of communicating accurate information, improving staff efficiency, or providing an effective working environment.

Publishing to a corporate website is normally fairly centralised. While material is drawn from various groups within the organisation, final review and publishing is typically managed directly by the web team (or equivalent).

This allows content accuracy to be assured, minimises legal risks, and coordinates with media releases and other public announcements.

With a much broader range of information and users, an intranet has content developed by many authors, following a decentralised approach. While a central team may still be responsible for oversight or final release, much of the content is directly published by the authors themselves.

These are just a few of the differences between websites and intranets. As each site improves in design and maturity, they will continue to grow apart. (A great website is very different to a great intranet.)

In practice, the same user-centred design methodology (Chapter 7) can be applied to websites and intranets. The fundamental principles of usability and information architecture also apply to both. Using these techniques when designing each site will, however, produce very difficult results and findings.

At a basic level, intranets and websites should look different. Staff need to be able to distinguish between website (publicly shared) and intranet (potentially confidential) information. The many other differences between the sites also mean that it doesn't make sense to simply copy the design of one site to the other.

Design also reflects the underlying role, purpose and management of the site. As will be explored in the coming chapters, there is a body of knowledge on how to deliver successful intranets that should inform design decisions. Intranets must be useful, and not just attractive and usable.

Differences between websites and intranets

Aspect	Public websites	Intranets
Business goals	Communicate information, support marketing, sell products	Broad goals, including: communicate information accurately, improve staff efficiency, support collaboration, streamline business processes
Audience	External users, wide range of skills and experience; limited understanding of organisation	Internal users, with a good understanding of organisation; wide range of information needs
Efficiency	Secondary issue for the site, unless frequently used by visitors	Primary goal of the site: to improve staff efficiency
Browsers & platforms	Many and varied	Consistent with a standard operating environment (SOE)
Size	Small to medium	Medium to extremely large
Content & structure	Narrow, structured around key products and services	Broad, varied information types and content
Content updates	Weekly or monthly	Daily
Presentation	Appearance very important for promotion and sales	Consistency more important than appearance
Publishing models	Often centralised, with the web team publishing most site content	Typically decentralised, with authors located in major business units across the organisation
Legal liability	Liable for every word published	Reduced legal exposure

Table 2-1: Summary of the differences between public-facing websites and staff-facing intranets.

What do we mean by design?

Many professional groups feel they own the word 'design', from artists to usability specialists. All have a useful role to play, and many elements go into delivering an intranet design that works well for staff and the organisation as a whole.

This book will take a holistic approach that brings together all the pieces to form a coherent whole.

'Design' is one of those words that means many things to many people. At a basic level, it is universally understood to be related to the visual appearance of things. In the context of intranets, this includes:

- colours

- logos and graphics

- overall branding and style

- page appearance

These are the domains of graphic artists, or visual artists. While these are some of the elements of successful intranets, they are perhaps not the most important.

In this book, we use the word design much more broadly, to encompass the whole intranet from an end user's perspective. This includes:

- intranet site structure

- navigation

- page layouts

- homepage design and content

These elements have the most direct impact on whether intranets are quick and easy to use. They also take intranet design beyond the scope of just a graphic designer, and introduce a new set of skills and experience.

This approach introduces the fields of 'usability' and 'information architecture' that provide a toolkit of techniques for creating designs that are easy and intuitive to use.

In the coming chapters, a structured methodology will be laid out that will address structure, navigation and page layouts. Smaller consideration will be given to the purely visual aspects of intranet design. (Although as discussed in Chapter 12, the intranet can't afford to be useful but ugly, and an attractive and engaging appearance is always important.)

What will not be covered is the underlying HTML involved in publishing intranet pages. This is an entirely separate topic, and we will not be covering how to actually 'build' the intranet. This is well covered by technical and web resources elsewhere.

What makes a great design?

Now that 'design' has been defined, it still leaves the question: what makes a great intranet design? There are a number of criteria that must be met, some relating to what is delivered for staff, while others relating to the value the intranet provides for the organisation it serves.

In practice, a great intranet design must meet the following criteria:

1. *Works well for staff.* The crucial measure of success for an intranet is that it meets staff needs, and is easy and satisfying to use.

2. *Useful for staff.* More than just being quick and easy for staff, it must also help them to be more productive in their day-to-day work. If the intranet isn't useful, there is little point in having one.

3. *Delivers business benefits.* Intranets are just a means to an end. To be considered successful, they must deliver concrete business benefits, in addition to meeting staff needs.

4. *Builds trust and confidence.* A great design sends a clear message to staff that the intranet is well maintained, well designed, and worth using. This includes creating a sense of emotional engagement with the site, and delivering content that is up to date and accurate.

5. *Reflects and reinforces organisational culture.* The intranet does not exist in isolation, and it needs to reflect the prevailing culture in the organisation, as well as being derived from overall corporate goals and brands.

6. *Improves sustainability and manageability.* The intranet team needs to manage the intranet over a long period. A great design makes this easier, for example by making it clear where new items should be added (and where they shouldn't).

Use these criteria as a checklist, to assess an existing design or to guide the design process. Balance must be found between the criteria, and no item can be overlooked.

For example, it is not enough to deliver an intranet that is easy and enjoyable to use, if it doesn't benefit the organisation. A 'perfect' intranet on the first day of go-live is worth little if it quickly becomes overwhelmed by the endless stream of changes and additions.

Over time, web practices evolve, and fashions change. While the intranet should not slavishly reflect the latest in design thinking, it does need to be modern enough to send the message to staff: "this is a well-maintained and professionally run site that works well". More on this in Chapter 12.

Why staff visit the intranet

To understand the importance of good intranet design, it is valuable to explore why staff visit the intranet. In practice, there are two fundamental reasons why staff visit the intranet:

- *To find a specific piece of information.* The staff member is looking for a specific fact, detail or figure, such as how much leave they have left to take this year.

- *To complete a specific task.* The staff member has a particular activity to complete that the intranet can help with, such as booking travel or applying for leave.

In both cases, the staff member is not looking for the HR manual, a procedure, or some other general resource. Instead, they are seeking something very specific to meet an immediate need.

It is also important to recognise why staff aren't coming to the intranet:

- *Not to check on the latest news.* Although this is a valid secondary use of the intranet, corporate news is typically not enough to draw staff to the intranet every day. The exception would be clearly operational news that is critical to daily work.

- *Not to browse around.* Staff are busy, and have little interest in exploring the uncharted corners of the intranet in their spare time. The exception are new starters in an organisation, who will use the intranet to build an understanding of who does what, and how. (Once they have gained sufficient experience, their usage reverts to match longer-serving staff.)

- *Not for fun.* While 'buy and swap' areas are popular, staff almost certainly prefer to have fun in their personal lives rather than on the intranet at work.

The intranet is most useful when it is directly useful for staff, helping them to find required information or to complete tasks. While this has a lot to do with the functionality of the site, the design of the intranet is also critical in helping staff successfully obtain what they want.

The results of the Worldwide Intranet Challenge (www.cibasolutions.com.au) reinforce these observations. The top three tasks listed by intranet users across the globe are 'find instructions for completing work tasks', 'upload or download documents' and 'complete online forms'.

Design and intranet strategy

Creating a great intranet design is just one element of creating a successful and valuable intranet. Equally important is delivering what staff need, having a clear overall strategy, and having strong ongoing management in place.

Intranets are only useful if they do what staff need. They can be wonderfully easy to use, but entirely useless if they fail to provide the required content and functionality. This is why researching staff needs is so important, as well as involving staff throughout the design process.

This is not a handbook on conducting user research, although it will be covered at a high level in Chapter 4. In parallel with the design process, ensure that careful consideration is given to the functionality that will be delivered on the new site, and make sure it is more than is currently provided to staff.

These decisions should be made in the context of the broader intranet strategy. This must define the overall purpose of the intranet, and the role it will play in meeting staff needs and organisational priorities. There also needs to be a clear roadmap for ongoing intranet improvements, ensuring that the site will deliver the hoped-for benefits. These topics are touched upon in Chapter 6, and will be explored in detail in a future book.

As with any large and content-rich site, intranets are not easy to sustain and grow. There are considerable management challenges, not least due to the widely decentralised nature of content authoring and ownership.

Intranet teams have been steadily building up a body of knowledge of what works, and what doesn't, when it comes to managing sites. This should be drawn upon to deliver a site that is sustainable, so that the effort involved in redesigning the site is not squandered.

One book cannot cover all of these topics; it would be the size of a telephone book! The focus here is on the design elements of the intranet, and guidance on some of the other aspects of intranets can be found in my other book, *What every intranet team should know*.

What every intranet team should know
James Robertson

Key elements of intranet strategy and management are covered in this deliberately concise volume, which highlights fundamental principles and models that are useful for all intranet teams.

For more information:
www.steptwo.com.au/products/everyteam

What kind of projects are we talking about?

There are as many different intranets as there are organisations, and intranet projects are similarly diverse. It is therefore vital to make clear the assumptions that underpin this book, and what types of projects it focuses on.

The first distinction is between 'greenfield' projects (the creation of a brand new intranet) and 'brownfield' projects (the redesign of an existing intranet). Table 2-2 lists some of the common differences between these projects.

Most medium to large organisations have had an intranet in place for some time now, often for a decade or more. As discussed in *What every intranet team should know*, these sites undergo a long period of organic growth before the imperative of good intranet design is realised.

Most intranet teams are therefore confronted with the task of redesigning an existing intranet (brownfield projects). Staff are generally unhappy with the intranets (otherwise why change them?), and the sites are large and sprawling.

While this does mean that a lot of work is required to clean up existing messes, the advantage of these projects is that there is something to build on. The good elements can be kept, and frustrations with the problems point the way to improvements.

This situation is the primary focus of the book, and the methodology outlined in Chapter 7 assumes a redesign. Tips, tricks and suggestions will be drawn from real-life organisations that have undergone a redesign.

In a greenfields situation, the organisation doesn't have an existing intranet, and one is being created from scratch. This is unusual, and many intranet teams may never have such an opportunity in their entire careers.

Starting with a blank sheet of paper gives the chance to create something genuinely new, and better. It can be hard, however, for organisations to visualise what they want without having an existing site as a reference point.

The approaches outlined in this book also apply to this situation, and where there are differences from a redesign they are highlighted at the relevant point.

It has also been assumed that the project is being conducted in a medium-sized organisation (up to a few thousand staff), located in a single country. This hits the 80/20 rule, covering the overwhelming majority of intranet projects.

Larger projects and more complex organisations have their own challenges, beyond a comparatively simple redesign. These differences are also covered throughout the book, and in Chapter 19 in particular.

Two types of intranet projects

Project	Benefits	Challenges
Brownfield projects (Redesigning an existing intranet)	Can build on what's already there. Strengths and weaknesses of current site can guide future designs. Need for an intranet already established (at least to some degree). Project can often be driven by frustration with existing site. Authors and content owners already established and involved.	Large and complex existing site(s). Considerable cleaning up and restructuring of current content. Ownership, cultural and political issues relating to the current design. Future desires are strongly influenced by current site functionality and designs. Harder to do a small project; goal is typically to fix 'everything'. Legacy sites and systems.
Greenfield projects (Creating a brand new intranet)	Fewer preconceptions about what the intranet should be. Opportunity to create something new, unencumbered by the past. New approaches can be tried (eg social intranets). Something small can be launched, and then built upon progressively. No existing content to manage or migrate. Can establish effective governance and working practices from the outset. Can establish a new technology platform.	Organisation and staff may struggle to identify what the intranet should do, in the absence of direct experience with intranets. Expectations that a new site can be created very quickly. Smaller budget and resources. Harder to demonstrate value and importance of a site that doesn't yet exist. No pre-existing content owners or stakeholders; these need to be established as part of project.

Table 2-2: Redesigning an existing intranet (the primary focus of this book) is very different from creating a brand new intranet where none has existed before.

Chapter 3

Common design mistakes

Intranet projects are challenging at the best of times. Intranets are large and content-rich. Project teams are often thrown into the deep end, with many constraints and expectations.

Intranet projects are often confronted with issues such as:

- unclear intranet ownership and governance

- tight timeframes

- limited (and often insufficient) budgets

- varied (and sometimes competing) stakeholder opinions

- internal politics

- large number of end users (staff), with widely varying needs

- technology considerations and constraints

- limited team experience and skills relating to intranets

- poor access to best practices and other intranets

Is it any wonder that intranet projects go off the rails? Even the most experienced and well-resourced teams can struggle under these circumstances.

It is therefore useful to explore common mistakes made on intranet projects, to provide context for the intranet methodology outlined later in the book. These mistakes will also highlight key principles for the design process.

Mistake 1: Not testing with staff

A lot of hard work is put into designing the new intranet. A number of options are considered and compared. Designs are carefully refined, tweaked and tuned. A home is found for every page, each in its logical place. The new intranet is easy, intuitive, and makes perfect sense.

Makes sense, that is, to the designers of the site. Which is to be completely expected, as they are the designers. They have lived every aspect of the project for months, and know every corner of the site. The real question is: does the new intranet make sense to staff?

This is the most common mistake made by project teams when delivering a new intranet. As designers, we are not users. The way we think is very different from that of 'normal' staff, and our expectations and practices do not match theirs. It is a tragedy when a huge amount of effort is put into delivering a new intranet, and it proves to be no easier for staff. Yet we have heard this many times.

The companion book *What every intranet team should know* outlines the six phases of intranet evolution.

Once intranet have passed through a stage of rapid organic growth, they can get trapped in an era of repeated redesigns.

The problems of the existing intranet force a redesign, which produces a site that is different in design and structure. As outlined in this first mistake, however, the intranet team doesn't test with actual staff.

The problems with the intranet remain, and eventually force yet another redesign. Some organisations have redesigned four or five times, each time delivering a site that is different, but no easier to use for staff.

The key omission was testing with actual end users: staff. There are a range of practical, hands-on techniques which can be used to evaluate designs and navigation, and to refine accordingly. These include card sorting (Chapter 9), tree testing (Chapter 11) and usability testing (Chapter 13).

Taking designs or mockups, putting them in front of users and asking "what do you think" does not count as testing. This gathers opinions, what people think about the design, but does not ensure that the design is actually easy to use when it comes to completing common tasks.

People's opinions about what they would do, or want to do, often do not match what they actually do in practice. (This is why many people talk about eating healthily, and yet fast food chains are some of the biggest retailers in the world.) Hands-on testing goes beyond this to assess more realistically whether the intranet is easy or hard to use.

Teams who make use of these techniques ensure that all the effort involved in an intranet redesign leads to the desired outcome: an intranet that really does work well for staff.

Mistake 2: Designing by opinion

There is no shortage of opinions on what should go in the intranet, and how it should be designed. These come from stakeholders, senior managers, staff and peers. One of the most difficult aspects of designing an intranet is to cut through these often competing opinions to produce an intranet that works well for staff.

Stakeholders are clearly an important group when designing an intranet, and projects can tend to rely on workshops and sessions with these managers when determining intranet strategies.

There are, however, considerable risks with this 'design by stakeholder' approach. First and foremost, they are not the actual users of the site. It can also lead to a publisher-centric approach to site design, driven by the needs of those who write the content, rather than those who use it.

Stakeholders know a lot about the business priorities of their unit, but they don't know a lot about intranets (nor do they need to!). While their input is valuable, it doesn't necessarily match what they do in practice.

At worst it can involve the CEO making design or colour decisions, or picking one of a number of options based on personal preference. While they are the head of the organisation, this doesn't make them an expert in intranets!

Intranet projects can therefore spend considerable effort delivering a site that meets the expectations of stakeholders and staff, without producing a site that is actually easy to use in practice. Staff complaints continue, and the intranet continues to struggle.

This doesn't mean that stakeholders should be shunned and staff ignored, quite the opposite. What is needed, however, is a structured way of engaging with the organisation that obtains the necessary involvement (and engagement) to produce a site that works.

The key role of stakeholders is to determine the overall priorities and focus for the intranet. At the outset of the project, this involves determining the overall 'intranet brand' (Chapter 5) which shapes design and functionality decisions. Stakeholders can then be kept in the loop as the research and testing activities are conducted, as outlined in Chapter 7.

Time must also be spent with content owners to help them understand the design process, and the rationale of the final design. Since they will be producing most of the pages on the site, their needs and considerations must also be addressed during the design process, but not in a way that derails delivering a site that works for end users (staff).

Mistake 3: Designing only half of the intranet

The intranet homepage is by far the most visible page on the site, and it gains the greatest traffic. It is not surprising that the most design attention is focused on this one page on the site.

The design process works down from there, typically concentrating on the top layer or two of the intranet. Below that, pages are left to business units and content owners to manage as they see fit. Unfortunately this is doing only half the job.

As outlined earlier, staff come to an intranet to find a specific piece of information, or to complete a task. This is considered successful when they have obtained what they need.

Having a well-designed intranet for the first few clicks means little if the staff member is then dumped in a poorly managed intranet section or site.

For example, it is not hard to work out that the leave form will be found in the HR section, making the first click easy to choose. If the HR section is very poorly designed, staff will still struggle greatly, just a single click into their search for the leave form.

Key sections of the intranet need to be designed all the way to the bottom of the site. This means applying good design principles and methodologies to content owned by decentralised business units, beyond just the top levels of the site.

Intranet teams also need to work closely with content owners, playing a mentoring and support role where appropriate, to help them deliver content that works well for a broad staff audience.

Mistake 4: Taking a publishers' perspective

Intranet content is owned by someone, such as HR, finance, IT, customer service or product development. These content owners are responsible for maintaining 'their' content on the intranet, whether it's one section or a whole 'site'.

It is human nature that content owners will create and publish content in a way that makes sense to them. They use terminology and jargon that is familiar to their business unit, and will structure content in a way that is logical to them.

This 'publisher-centric' approach to intranets is widespread and corrosive. It means that staff have to know who owns the content before they can find it, one of the most common frustrations for users, particularly in large and complex organisations.

It leads to intranets that are dense in jargon, where whole sections of the intranet are named in surprising ways. For example, staff all understand who 'HR' are, and what they do. These areas, however, have a tendency to rename themselves and then use that as the navigation on the intranet ('People & Performance' anyone?).

At the worst extreme, this produces intranets which are little more than a collection of independent sites, islands of content in a common ocean. This is commonly the case for very large organisations, where business units operate independently and may be the size of medium-sized firms in their own right.

The challenge for intranet designers is to move the site towards a 'user-centric' approach, where content and functionality is delivered in a way that will be easily understood by staff.

In one very large organisation, the homepage of the intranet has the following links:

- Ask HR
- People system/ePerformance
- eOrganisation
- Overtime
- Better people manager
- Leave (eLeave)
- Recognition
- Benefits plus
- Payslips
- Health and wellbeing

This would be a baffling array of links for the general staff person, particularly if they don't understand the distinctions between the services being offered.

These problems are the result of a publisher-centric view, as well as being the by-product of ongoing organic evolution.

Mistake 5: Allowing technology to rule the design

Intranet technology is constantly evolving and improving, as evidenced by the current spread of SharePoint and wiki-based intranets. New tools offer greater capabilities and new approaches that can benefit organisations and staff. While technology improvements are always appreciated, they also lead intranet and project teams into common traps.

The first is to assume that more functionality is better than less. With any new tool, the temptation is to enable most (or perhaps all) of its features. This can, however, quickly overwhelm the readiness and skills of authors, content owners and staff (end users). With greater functionality also comes more complexity, with the requirement for additional training and support.

There can also be an assumption that if a capability is offered in a product, particularly a widely used solution, then it must work in all situations. This overlooks the importance of understanding staff needs, matching the organisational culture, and setting a manageable pace of change.

It takes a lot of time and effort to 'digest' a new technology solution, understand how it works, and become proficient in its use. For this reason, intranet redesign projects can all too easily become technology projects, with the IT aspects overwhelming the more important goals of improving the site itself.

Changing technologies also forces a complete migration of content from the old site to the new platform. In turn, this triggers a 'big-bang' redesign, with all the effort that entails. This eliminates the possibility of incremental improvements, or step-wise changes.

Of course, if the current technology platform for the intranet is completely broken, then a new product will be very welcome, and likely a necessity. Technology enhancements also help to drive the steady pace of intranet evolution. Care must be taken, however, to ensure that the benefits of the new tools are maximised, while the impact is reduced or mitigated.

Mistake 6: Copying intranet designs

Jakob Nielsen of Nielsen Norman Group (*www.useit.com*) once compared the homepages of a number of intranets, overlaying them to show that they all have a very similar design. This is useful to know, but it is more important to ask: are these designs any good? Or are the intranets all the same, and equally wrong?

The widespread nature of staff complaints about intranets suggests that we have some way to go before we consistently deliver sites that work well. With many organisations falling into the traps outlined in this chapter, the danger is that copying another site may simply copy their mistakes.

It *is* useful to see other intranets, and to take from them ideas and design elements. Care must be taken, however, not to simply replicate the 'way things are done' without checking that it will work well for *your* staff (see mistake 1).

Intranets work best when they reflect the organisations they serve. Every organisation is different, based on its purpose, size, history, culture, geographic spread and a hundred other factors. For this reason, there is no single 'best' intranet design, only the design that best meets the needs of staff in your organisation.

Project teams also need to be wary of common 'rules' and 'principles' which turn out to be myths. The 'three clicks rule' is the most widespread example of this (see Chapter 10 for more on this). Just because something is done widely, doesn't mean it's right.

Instead of these simple, but often incorrect rules, teams get the best result following a user-centred methodology that tests new designs with actual users.

Mistake 7: Designing in a hurry

Intranet projects can be hard to get off the ground, and significant time is spent building support, gathering resources and obtaining budgets. Once final approval is gained there is significant pressure to deliver a solution quickly. With the intranet team talking about the project for over a year, there is little patience for a drawn-out redesign.

This forces too many teams into a rapid redesign, completed in a small number of months, or even in a handful of weeks. Once trapped into 'designing in a hurry', intranet teams are forced to make the project activities fit the overriding time constraints.

This compromises several key aspects of the project. It limits the time and resources that can be spent on understanding the current intranet, and exploring staff needs. The design process itself needs to be cut short, potentially preventing any meaningful testing with staff.

Opportunities for meaningful engagement with stakeholders is limited, leading to potential issues down the track. It also leaves the project vulnerable to arbitrary decisions, by team members or senior management.

There is a limited time to review and correct content, leading to a straight migration to the new site. This 'garbage in, garbage out' approach produces a site that may be better structured, but still filled with out-of-date, incorrect or incomplete information. While the intention may be to come back to the content and clean it up after the initial launch, human nature means this is unlikely to happen.

These short-cut projects can come about when a change in technology platform is driving the redesign. As the technology deployment slips, it eats into the remaining time that was allocated to the redesign itself.

There is little value in relaunching an intranet if it's not better. Even short projects take a lot of effort, and will undoubtedly generate a lot of stress on the project team. If the new site is not clearly better, the value of the redesign will certainly be questioned.

A full-scale intranet redesign will not take any less than nine months, and 18 months is common. (Any longer than that, and the project has lost its way or has attempted to do too much.) Expectations around these timeframes need to be set at the outset, and constantly communicated to stakeholders.

This is not to say that all redesign projects need to be this large, and there are many benefits to taking accelerated or incremental approaches to the redesign, which are discussed in Chapters 16 and 19.

Mistake 8: Not designing for the long term

Assuming the design or redesign project went well, the intranet looks great on the day it launches. The layout is elegant and attractive, navigation is intuitive, and every piece of content has its place on the site.

Now the challenges really start. From the day the site is launched (or relaunched), requests start rolling from the business. This includes new content to be added, entirely new sections, content updates and additional functionality.

Everyone also wants their content, news or project on the homepage. Requests travel up the chain of command, until a senior manager requires the item to be added to the homepage.

Very quickly, the clean design and intuitive structure can start to slide. Unless managed well, the intranet will quickly become a mess. This is like cleaning up the attic: it's perfect one day, but fast forward two years and it's filled again with junk.

Intranet teams can overlook the longer-term sustainability and manageability of the site when developing a design. Insufficient space is left for additional items. Page elements have unclear purposes, and new items don't have a clear home.

These sustainability considerations sit alongside the ease of use for staff, but they are equally important. A well considered design will greatly help the intranet team to manage the ongoing updates and changes to the site, as well as somewhat mitigating the impact of internal politics and opinions.

Mistake 9: Stopping at the end of the project

Intranet projects are too often run as one-off initiatives to 'fix' the intranet. The objective is to launch or relaunch the intranet, with a new design, structure, navigation and features. The project team works hard for six, 12 or even 18 months, leading up to the 'go-live' date.

Once the intranet is launched, the site goes into 'maintenance mode' or 'business as usual'. Content is published and updated, but no significant changes are made to the intranet once the project has been completed.

There are two major problems with this approach.

The first is that a single project, no matter how large and well-resourced, cannot address every issue or deliver every desired feature. Every project is limited by constraints, leaving many needs unmet.

Intranet needs also evolve over time, in response to changing business requirements, new technologies, and staff issues. Without ongoing work, intranets immediately start to fall behind business needs the moment they are launched.

Within two or three years, intranets are often looking dated again, and staff complaints have risen again to pre-redesign levels. This often triggers another redesign, and the cycle starts again.

Successful intranets are delivered by an ongoing 'process', rather than a one-off 'project'. While an intensive project may initially be needed to address big issues, this effort is for nothing if it's not supported by a continual process of improvement and refinement.

Outcomes

Intranet projects are often challenging, and hidden away within organisations, it is not surprising that teams fall into common traps. Failing to test with users, or designing by stakeholder opinion, impact heavily on the ease of use of the final site. Limited time and a focus on just the top levels of the site also reduce the benefits that are delivered.

Many of these issues can be easily addressed when a robust methodology is followed. Even within tight time and resource constraints, a simplified approach can still be taken that delivers many of the same benefits. This methodology will be outlined in the chapters to come.

Chapter 4

Understand staff needs

A good intranet is one that works well for staff. To achieve this, designers need to have a strong understanding of staff needs, the environment they work in, and their daily tasks.

In most organisations there will be a hugely diverse range of staff, which may include:

- managers (senior and middle)

- admin staff and other back-office workers

- front-line staff (eg call centre, sales, customer service)

- IT teams and other corporate services areas

- professionals (eg lawyers, consultants, auditors)

- specialists (eg product experts, researchers)

- staff working 'on the floor' (eg nurses, doctors, factory workers)

- staff working in the field (eg engineers, inspectors)

- staff in remote or regional areas

(This is not an exhaustive list, and staff will often fall into multiple categories.)

These staff are not interchangeable, and they have very different requirements and expectations for the intranet. In order to deliver an intranet that works for them, time must be spent researching their specific needs. This forms a key input into the design process.

Start with research

The starting point for any design or redesign project is to conduct research with end users. In the case of intranets, this involves spending time with staff throughout the organisation.

There are clear objectives for this initial research:

- identify strengths and weaknesses of the current site (if there is one)

- understand current usage levels and patterns

- identify major staff groups and key audiences

- uncover staff frustrations, bottlenecks and points of pain

- understand staff activities and needs

- identify opportunities that can be met by the intranet

- define the focus of the design or redesign project

The overarching outcome from the research is to determine what to deliver on the intranet, and how. Information about staff needs will shape every aspect of the intranet design, from the functionality of the homepage down to content pages within each section.

An understanding of working practices and organisational culture is also the starting point for defining the overall intranet strategy, and the 'brand' of the intranet (which will be discussed in Chapter 5).

Don't ask staff what they want!

To uncover staff needs, it may seem sensible to ask questions such as:

- What is most useful on the intranet?

- What areas of the site do you use most often?

- What are the problems with the current intranet?

- What features are missing from the intranet?

- How could the intranet be improved?

- What additional information do you need?

- How else could the intranet help you with your job?

While they seem reasonable, these questions are almost entirely useless, as they require staff to have an understanding of intranets, and how they can support day-to-day work in organisations.

Staff have little understanding of intranets, because they don't need to. Instead, they have in-depth knowledge about their own role and activities, and it's the job of the intranet team to understand intranets.

Asking these types of questions generates one of two possible responses:

- "I'm not sure. Can you give me some examples of how an intranet could help me?"

- "I think it would be great if the intranet provided feature xyz!"

Is it time to get some specialist assistance? All of the techniques outlined in this chapter are usable by every intranet team. With some basic skills in good interviewing technique, an in-house team can quickly uncover key information to help guide the design process.

On the other hand, there can be benefits in gaining external, specialist assistance when conducting user research.

Coming in with a fresh perspective, consultants or contractors can often uncover issues that have been taken for granted within the organisation. They can also bring in-depth knowledge of both user research and intranets to bear.

If external staff are used, time will need to be allocated to transferring the knowledge and findings to internal team members, as this information will be of use in day-to-day work, and not just for the project.

In the first case, staff are unable to provide meaningful input into intranet design. This should not be a surprise, as staff are not expected to be intranet experts. Instead, they are hired to be very good at their specific jobs, and need to know little about intranets beyond how to use them for their day-to-day work.

In the second case, a wish list of features and tools is collected, but without ensuring that these ideas will actually be useful (or used) in practice. These requirements can also be driven by current fads for new technologies and approaches. The wish list can even be dangerous, creating expectations that the new intranet will deliver these features, even when they aren't needed.

Staff will generally respond based on their current use of the intranet, which will reinforce existing practices, and make it harder (not easier) to build an innovative intranet. When a new intranet is being created for the first time, the absence of an existing site will also make it hard for staff (and the organisation as a whole) to articulate intranet requirements.

In short, attempting to gain intranet ideas from staff in this way will not generate the hoped-for insights. Instead, focus on what staff do and how they do it, as listed on the following page.

Intranet teams should take a structured approach to the staff research, using a mix of practical techniques to uncover needs and issues.

Ask the right questions

The previous pages have highlighted the problems with asking staff what the intranet should do. But what are the right questions to ask?

When talking to staff, focus on building an understanding of how they work, and what they need in a day-to-day sense. This might include questions such as:

- What do you do in your job?
- What does this involve on a day-to-day basis?
- What information do you need to do your job?
- Where do you currently get this from?
- How do you find out about what's happening in the organisation?
- Who else do you work with in the organisation?
- How do you keep in touch and collaborate with them?
- What are some of the frustrating or difficult tasks you've had to do in the last few weeks?

This should not be considered a 'clipboard carrying' exercise of filling in the blanks in a questionnaire, and these questions are just starting points for the discussions with staff.

The fundamental goal is to build a holistic understanding of:

- how staff work
- what they do
- the environment they work in
- any points of pain, unmet needs and opportunities for improvements

To achieve this, follow a 'semi-structured' approach. This involves starting with open-ended questions, and then exploring topics in depth as they arise.

For example, the 'policy & procedures manual' could come up in discussions. Further questions would then uncover what this is, where it's located, how it's updated, and the way updates are notified to front-line staff.

Look for patterns across the research, such as key issues that impact every staff member, or common needs that surface for a number of job roles. Conversely, the research may uncover requirements that are perceived as vital by one group, but are not relevant elsewhere, or cannot be practically met by the intranet.

Taking this approach will always uncover surprising findings, including intranet opportunities that were not previously recognised. It will also give vital context for the design process, as well as identifying important issues relating to work practices and organisational culture.

Use structured techniques

Intranet teams can easily uncover true staff needs without ever asking 'what do you want?'. Instead, they can use a range of more structured ways for uncovering needs.

These needs analysis techniques include:

1. *Surveys.* These are commonly used to gather the input of staff throughout the organisation, and their big advantage is the ability to obtain a large number of responses. Their tendency to fall into the trap of 'asking staff what they want', however, means their use should be restricted primarily to assessing staff satisfaction with the current site, rather than as the main form of user research.

2. *Focus groups and requirements workshops.* These are facilitated discussions that focus on exploring topics within a group setting. Often used as a way of gathering input from larger groups of staff, they are best used to explore current issues and problems, rather than to discuss 'wish-lists' of potential intranet ideas.

3. *Staff interviews.* One-on-one interviews are a very effective way of gathering information on staff needs and issues. When conducting interviews, avoid asking questions about the intranet itself. Instead, take a semi-structured approach to exploring how staff work, what this involves day-to-day, the information they need, and where they currently get this from.

4. *Stakeholder interviews.* A variation on staff interviews, stakeholder sessions focus on middle and senior managers. The goal of these sessions is to understand business drivers and strategic goals, as well as business-area considerations relating to the intranet.

5. *Workplace observation.* This involves going out into the field, to observe the activities of staff and the environments they work in. Workplace observation is particularly effective in finding out what staff need in call centres, manufacturing areas, field working, or on the road.

6. *Contextual inquiry.* A mix between interviews and workplace observation, this involves sitting with staff in their working environments. In addition to answering questions about day-to-day activities, staff members show how they complete common tasks. This provides in-depth insight into how staff interact with and use key systems and sites.

Together, these techniques will quickly build up a comprehensive picture of the organisation, and where the intranet can be of use. They are also holistic, building up insight into staff culture and working practices.

These insights will be of benefit to other projects within the organisation, including internal communications, document and records management, business process re-engineering, to name just a few.

> Step Two has written extensively on how to make use of these structured approaches to understanding staff needs. A good starting point is the following article:
>
> www.steptwo.com.au/papers/kmc_needsanalysis

Bring together every source of information

Intranet projects are not conducted in a vacuum, and there will be considerable work already done in other areas of the organisation that can be used as part of the design process.

This includes:

- usage statistics

- search engine usage (such as most popular searches)

- staff satisfaction surveys

- internal communications research

- corporate strategy documents

- IT strategy documents

- chatting to people over coffee

- participating in project meetings

Avoid reinventing the wheel by gathering together all available information, and use this to supplement the findings from the structured needs analysis outlined earlier.

This will also help to ensure that the intranet design work is aligned with other projects and strategies within the organisation. This reduces overlap or conflict, and mitigates potential internal politics that can arise from competing projects.

Case study: redesigning a call centre intranet

Conducting needs analysis can often overturn assumptions about intranet design and project priorities, as well as surfacing previously unknown requirements. Take a call centre within the financial services industry as an example.

Call centres are high pressure environments. With potentially hundreds of staff all taking calls, very close attention is paid to productivity and throughput. This includes an illuminated display on the wall showing the number of calls taken so far, how this compares to call centre targets, average call handling times, and more.

It is well recognised that call centre staff require a huge body of knowledge, and that this changes frequently. Less understood is how this information is typically disseminated to the call centre.

The primary way of delivering updates to the call centre is via email. These are sent from multiple sources in corporate head office, including marketing, sales, product development and legal. They may be sent just to the head of the call centre, or to individual staff, or both.

Staff carefully file away these emails in elaborate folder structures in their email software, sometimes forwarding emails back to themselves so they can add a better title.

This particular call centre had an existing intranet, although it was in a very poor shape. When asked, staff indicated that they didn't use it. They were aware of the site, a major promotion had been done around it some months earlier, but it wasn't being used.

When asked why, the universal answer was: "if it's important, it's emailed to me". They were right. Everything arrived via email, but only some things were placed on the intranet. So if they needed to find something, email was the most trustworthy source.

This was the first important finding: as long as email was used to disseminate updates, the intranet would never prosper. Emails needed to be limited or eliminated, with the information moved to the intranet. To be trusted and used, the intranet would need to contain a complete set of updates, and an easy mechanism to find earlier information.

When prioritising which content to focus on, there is a natural inclination to start with the most frequently used, and therefore most valuable content. In this case, it was details on the organisation's financial products.

These are contained in brochures called 'product disclosure statements' (PDS) that contain the high-level information plus all the fine print. These are provided to customers before they sign up for a new account.

Every staff member in the call centre has copies of the PDS booklets on their desk, and it turned out that the printed documents were the best solution. When customers rang up, they would often ask "on page 5 it says x, what does this mean?". The customer service representative could then quickly pull out the correct PDS, flip to page 5, and explain the details. This was much quicker than any electronic format.

So the most recent and most used information was best left in paper. Instead, it was the old information that most needed to be put online. In the financial services industry, customers open accounts and then can hold them for 20, 30 or 50 years in the case of retirement accounts. The accounts are still bound by the terms put in place when they were first opened.

When a customer rings up to close a 40-year-old account, what are the conditions and costs? The existing solution was to rifle through a cupboard in the middle of the call centre to try to find the old contract, with the customer still on hold.

The solution here was to create an online archive of all the PDSs, brochures and contracts going back as far as they could. Even as a simple scanned PDF, these would be much more easily found and used than a cupboard filled with old documents.

If time hadn't been spent in the call centre, none of these findings (and more) would have been known. A new intranet could easily have been designed with the best intentions, but based on incorrect assumptions, may never have been used.

Plan the research

A large organisation will have thousands of staff, and it is clearly not practical to talk to them all. Even in a smaller organisation, the thought of trying to talk to sufficient numbers of staff can be daunting.

Time and budgets tend to be tight in intranet projects, even in larger organisations. Once the project starts, there can also be pressure to 'get into the design' and to deliver something concrete quickly. All of which can make it hard to find sufficient time to conduct thorough needs analysis with staff.

Thankfully the focus of the needs analysis is to uncover key findings and major issues. There is no requirement to uncover every issue, as there can never be sufficient time or budget to meet all needs!

For this reason, even a week's time spent talking with staff and observing their work will be valuable. This will uncover sufficient information to underpin a 6–12 month project.

There are two main ways of selecting staff to include in the needs analysis, shown in Figure 4-1.

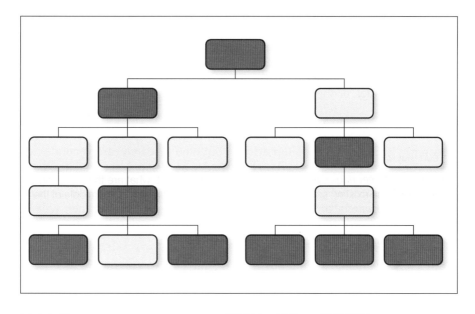

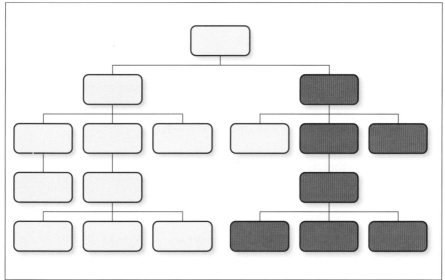

Figure 4-1: Two approaches to selecting staff to include in needs analysis. Taking a horizontal slice (above) gives a broad perspective on intranet needs, while a vertical slice (below) informs detailed design decisions.

- A *horizontal slice* aims to include a representative cross-section of staff from across the organisation. This will uncover broad needs, and will inform the design of the top levels of the site.

- A *vertical slice* conducts in-depth research with a specific business area or group of staff. This supports the design or redesign of a targeted section of the intranet, or the creation of a new piece of functionality.

In practice, a mix-and-match approach can be taken. The intranet team can start with a broad cross-section of staff and then follow up with more in-depth research with key groups. Alternatively, targeted research can be conducted as part of a pilot project, backed up by broader needs analysis at a later date.

If time or resources are particularly tight, do not despair. Even a small amount of informal research, or time spent watching staff use the intranet, will be valuable. Additional 'guerilla' research can also be conducted as the project unfolds, gaining key information at the point of need.

It can also be valuable to gain outside assistance when conducting the needs analysis. This draws on the experience of the consultants, and can help to accelerate the overall project timetable (as long as the budget supports it).

Focus on key staff

Not all staff are equally important. As highlighted earlier, it is not practical (or desirable) to attempt to include all staff across the whole organisation. Even when a horizontal or vertical slice is taken, this can still leave many potential staff to cover.

Research should therefore focus on the staff who are most valuable in the organisation, and most relevant in the context of the intranet. When selecting staff, consider choosing:

- *Frontline and operational staff.* In most organisations, the most important staff are the ones who deliver the actual services, or create the products. These staff can often be found in frontline or operational environments, such as in call centres, factory floors, on the road, or in hospital wards.

When designing intranets, the golden rule is this: you can't deliver effective solutions to staff you haven't personally met.

In an organisation of 1,000 or 10,000 staff, this doesn't mean that the intranet team needs to meet them all.

It does mean, however, that if the goal is to support particular staff in their jobs, it is critical to understand what they do, how they do it, and the environment in which they work.

Without this information, there is every chance that well-meaning projects and designs will miss the mark, delivering much less (or no) value.

It is therefore critical that intranet teams spend time at the outset of the project understanding staff needs. Even when done informally, this will greatly increase the chances of project success.

- *Key business areas.* Service delivery and front-office areas of the organisation are typically more important than back-office or supporting business units. By focusing on areas that generate revenue, the project team will uncover ways that the intranet can directly support business objectives.

- *Information consumers rather than producers.* Intranets can easily become structured according to the needs and interests of those who publish the content. To escape this, staff research should focus on information consumers, to understand their real (rather than perceived) needs.

- *Users rather than stakeholders.* Intranets should be designed for use. While stakeholders have a valuable role to play in setting strategy and managing content, they are often not actual users of the site. A clear line should therefore be drawn between end users (staff) and stakeholders when conducting research.

- *Remote rather than central staff.* Staff in head office have many sources of information, including face-to-face meetings and informal discussions. Staff in regional and remote areas are much more dependent on formal sources of information (including the intranet), and are often beset by a range of practical challenges such as low bandwidth.

Spending time with staff is a unique opportunity to find ways of making the intranet more useful and valuable. Concentrate on the hands-on staff in key business areas to ensure that the maximum value is gained.

Identify staff tasks

Intranets are at their most valuable when they help staff complete common tasks. This includes finding a key piece of information ('what is the address for our interstate office?') or completing a process online ('I need to apply for some leave over Christmas').

A key outcome of understanding staff needs is a list of tasks that the intranet should support. These include both common tasks, and tasks that are corporately important.

There are many ways of uncovering staff tasks:

- *Interviews and focus groups.* Discussions with staff, whether one-on-one or in a group, will naturally uncover information about key tasks. Staff will also express frustration with current bottlenecks, as well as highlighting their most important activities. Guide these discussions to ensure that sufficient information is gained about intranet-based tasks.

- *Hands-on activities.* In addition to uncovering information about tasks as a by-product of broader research, specific activities can be conducted that allow staff to list and prioritise their activities. These can be done individually or in groups.

- *Usage statistics.* Examining the most popular pages on the intranet will help to infer common staff tasks. Note that these will primarily focus on finding information (rather than completing processes), and may be biased by the idiosyncrasies of the current site design.

- *Search engine usage.* When a staff member types 'leave form' into the search engine, we can be confident that they are looking for exactly that. Common search terms are therefore an extremely effective way of uncovering common staff tasks.

By the end of the user research phase, the project team needs to have a concrete list of tasks, at least in draft form. In addition to informing the overall design process, these tasks underpin tree testing (Chapter 11) and usability testing (Chapter 13).

Note that card sorting (Chapter 9) will also uncover information about staff tasks, and the tasks will be progressively refined as they are used through the design project.

Consider using personas

One of the greatest challenges is ensuring that the insights gained from the needs analysis are fully used throughout the design process. Too easily, staff needs are overlooked, or lost amongst the opinions of stakeholders and designers.

An effective way of capturing and communicating information about staff is through the use of 'personas'. Personas are archetypal users of an intranet or website that represent the needs of larger groups of users, in terms of their goals and personal characteristics. They act as 'stand-ins' for real users and help guide decisions about functionality and design.

Personas identify the user motivations, expectations and goals responsible for driving online behavior, and bring users to life by giving them names, personalities and a photo.

Note that this is different from traditional 'audience segmentation' done by market researchers, which tends to focus on demographics such as age, gender, etc. While this is useful in a broader context, personas provide more indepth information on key staff groups to inform the design process.

Feng *the infrequent user*

"Do we have to know this in order to pass?"

- Inexperienced library user
- Only interested in what's required
- Research is not a key part of course

Library usage

How often: weekly
For: study, meet friends
Advanced features: rarely
Reliance on MUL: high (few alternatives)
Material: books
Uses: catalogue
Comes into library: daily

Personal information

Age: 20
Profession: full-time student
Field: double, Finance and Accounting
Home life: single, lives in dorm
Hobbies: hanging out with friends
Personality: arrogant, ambitious

Computer usage

Experience: high
Primary uses: IM, email, web, Word
Favourite sites: BRW, Hotmail, eBay, YouTube
Hours online per week: 40
Works out of: library or home
Computer: laptop

Goals and motivators

Feng uses the library to...

- study between classes
- meet with friends
- get assignments done

Library's objectives

We want Feng to...

- meet her immediate needs
- learn better practices
- come back for more
- undertake training

Profile outline

Feng doesn't use the library for much other than a desk to sit at.

His course is more about facts and figures; the library is too old fashioned.

He considered himself expert at using computers when he started uni, but quickly found he had no experience with the programs the library has. They're really hard to use.

Sometimes Feng will find a book to read in the library if he doesn't have his textbook. But usually they don't are a few editions behind.

When Feng does need to look something up for an assignment, he'll ask his friends and see what they found. It's not worth spending time looking too far, he'll just take the first thing that looks ok.

Clients like Feng are...

- ELS undergraduates
- Law undergraduates
- ICS undergraduates
- predominantly male

Figure 4-2: One of the personas created to represent the needs of library users within a university. Material courtesy of Macquarie University Library.

Personas are most relevant when there are clearly defined groups of staff with distinct needs and environments. For example:

- staff in call centres

- sales people out on the road

- nurses in wards

- workers on the factory floor

- engineers working on rail lines

Figures 4-2 and 4-3 show two examples of personas in these types of situations. They provide a rich and engaging description of staff working practices and needs, and have proved to be very effective in guiding decisions.

Personas are less useful when it is hard to draw clear lines between staff. For example, administrative staff working in head office may not be distinct enough from policy officers or HR staff.

To create personas, review all the research data and look for patterns in attitudes and behaviors. For example, there may be staff who need to access information under strict time pressures, staff who spend a large amount of their time researching, and those who like to be seen as the experts in the organisation.

While reviewing the patterns, start to cluster the findings into individual personas. Give each persona a brief description, such as 'in-house expert' or 'travelling salesman'. While there is no ideal number of personas, try to keep the set small. Four or five personas work as effective design tools, while more may introduce the same confusion as a large user requirements document.

For further information on creating personas, see the following introductory article:

www.steptwo.com.au/papers/kmc_personas

Case study: personas at the Environment Agency

The Environment Agency in the UK is a government agency whose aim is to protect and improve the environment. 4,000 of the staff are 'front line', meaning that they are largely field-based and return to the office to do admin work. The remainder are office- or laboratory-based.

The intranet at the Environment Agency had grown organically, and like many such sites, lacked focus and coherence. In preparation for a major intranet redesign, a persona pack was developed to articulate the key audiences.

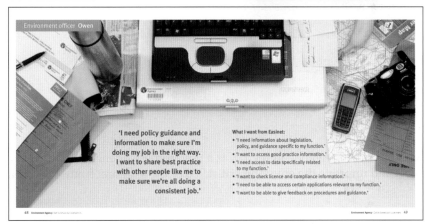

Figure 4-3: Award-winning personas pack created by the Environment Agency to support a major redesign. Samples courtesy of Environment Agency.

The personas pack won a Gold Award in the 2007 Intranet Innovation Awards (*www.steptwo.com.au/products/iia2007*), and three of the personas focused specifically on the intranet:

- Nikki – New Starter (her goals are common to almost all employees but especially to staff new to the organisation)

- Marcia – Manager (she needs information on how to manage staff, budgets and planning)

- Owen – Environment Officer (representing front line staff, who need specific process/procedural or legislative information to do their job)

As shown in Figure 4-3, a short booklet was developed that presented tasks and goals from the persona perspective. The 'put yourself in your customer's seat' tag was used throughout. This was backed up with strong visuals that showed the personas' work desks and described their key tasks and goals.

A set of summary cards was also produced, along with a mouse mat showing all personas.

The persona packs are being used to:

- commission new content and services

- provide guidance on what our audiences want and how they want it

- deliver content that is better written and more usable

- help authors to be more focused in the content they are developing

- identify new 'killer apps' for the intranet

- reduce the vanity publishing and content duplication

More broadly, the publishing teams are using the examples in the personas to sharpen their analysis and to question whether new content 'is really the best way of doing this? Is that what your customer needs or wants?'

The User Is Always Right
by Steve Mulder and Ziv Yaar

This is a very practical book on creating personas for websites and intranets. It provides an in-depth methodology, with three different options depending on time and resource constraints. Highly recommended.

ISBN: 0321434536

Business case and measurement

The primary goal of the user research is to understand how staff work, and where the key opportunities exist for the new intranet. This leads directly into the design process.

In many cases, it may also be necessary for the project team to demonstrate the need for a redesign project. This may include developing a formal business case, potentially quantifying the benefits that will be delivered.

There are a number of ways that the research can underpin an intranet business case:

- *Targeted activities.* Improving the design and structure of an intranet can often be an abstract task, difficult to explain to those not familiar with usability principles. The staff research should highlight key opportunities and issues that the intranet can assist with, making the intranet redesign into a concrete, business-focused activity.

- *Quotes and stories.* Direct quotes can be gathered anonymously from staff when conducting interviews, and these can very powerfully support recommendations and findings. Storytelling also helps to transform theoretical issues into concrete issues that impact directly on staff and the organisation. (Don't be afraid to use emotion when creating a business case.)

- *Videos and examples.* Where appropriate, video interviews with staff can clearly articulate dissatisfaction with the current site, and describe desired improvements. Photos, examples of problematic documents, and other samples can also flesh out the intranet business case.

- *Usability testing.* As will be covered in Chapter 13, usability testing is a practical way of checking that the intranet can be successfully used by staff. In addition to helping refine new designs, usability testing can also demonstrate the clear problems with the existing site. Inexpensive tools can be used to record simple videos (such as the one shown in Figure 4-4), which can be compelling watching for content owners and stakeholders.

There can also be significant benefits in conducting before-and-after measurements of the intranet. These can be used to show in concrete terms how the relaunched intranet is an improvement.

If senior management or project methodologies require the intranet team to demonstrate return on investment (ROI), this typically requires measurement before and after the project.

There are a number of ways of conducting this measurement, depending on the resources available, and the clarity around the role of the intranet in the organisation. These include:

- *Satisfaction surveys.* While surveys provide limited information to guide the design process, they are a well-tested way of assessing staff satisfaction. Either as a stand-alone survey or as part of a broader communications or staff survey, it is straightforward to determine satisfaction with the current site. Running another survey after the redesign should then show clear improvements. (Although note the change management challenges outlined in Chapter 15.)

- *Business measures.* The strongest case for intranets can be made when they directly address business issues, or support the delivery of broader business benefits. This could include improving customer satisfaction, reducing transaction costs, reducing error rates, or saving money. In these cases, initial research can quantify both the opportunities for improvement, and the potential size of benefits.

- *Task completion.* Benchmark tests can be conducted before the redesign, using techniques such as tree testing (Chapter 11) or usability testing (Chapter 13). These may show, for example, that only 54 per cent of tasks are successfully completed before the redesign. After the project, this figure should be much higher, and retesting will quantify this.

- *Time saving.* When the intranet redesign focuses on improving tasks, it becomes possible to determine potential time savings. Initial measurements can be taken of how long it takes to complete key tasks, and these can be reassessed on the relaunched site. Care should be taken when scaling this up to the whole organisation, particularly when endeavouring to put a financial figure to the improvements.

In practice, it can be hard to find time and support for extensive preliminary work before starting into the actual design process. Even when time is limited, however, it should be possible to conduct required research in a way that will generate at least some additional information for use in a business case or for demonstrating benefits afterwards.

Note that the stronger the requirement for demonstrating benefits, the more carefully the design project needs to be planned from the outset. It is not typically possible to justify benefits after the fact, if activities and outcomes have not been specifically chosen at the start of the project to support measurement.

The good news is that once benefits have been strongly demonstrated, it should become easier to gain support for future projects.

Figure 4-4: Usability testing videos can provide compelling (and engaging) evidence of the difficulty staff have with the current site. Screenshot courtesy of Lotterywest.

Consider remote research

In large and geographically dispersed organisations, it can be hard to conduct all staff research in person. (While it would be great to fly to Hawaii to visit staff in a regional office, this might be difficult to get signed off!)

While face-to-face research will always uncover the greatest depth of information, there are a range of practical options for conducting remote research:

- *Phone interviews.* The lack of non-verbal interaction makes these harder than in-person interviews, but they are still workable and useful.

- *Online interviews.* The use of screen-sharing and tools such as Skype allow remote staff members to walk through what they use on the intranet, and where their frustrations lie.

- *Diaries and activities.* Staff can be asked to fill out diaries outlining their activities, or can be mailed physical copies of card sorting cards (Chapter 9) or other activities.

Use these techniques where necessary to supplement detailed research conducted with local staff.

Conduct informal research

Intranet teams gain the greatest insight by following a structured approach to understanding staff needs. This involves dedicating enough time to acquire sufficient information, and using a mix of research techniques.

Intranet projects, however, invariably have to work within constraints. If there is limited time, resources or budget, it may be difficult to conduct full-blown user research.

In these situations, even a little research is better than none. Interviews can be held informally with staff over a cup of coffee. When visiting other sites or participating in meetings, a little extra time can be squeezed in for discussions.

Folding the research into other activities can also be a good way of learning about staff needs without having a stand-alone item in the project plan. This is the best chance to identify opportunities to add value via the intranet, and teams should benefit from this wherever they can.

Teams should also establish an ongoing culture of research, even after the design project. Insight can be gained incrementally, and then used to guide further site improvements.

Outcomes

Understanding staff needs provides a foundation for the entire intranet project. By the end of this initial phase of activity, intranet teams will have identified:

- usage patterns on the current intranet

- problems and issues with the existing site

- key staff needs and tasks

- opportunities for the intranet to better assist staff

- key cultural factors influencing intranet usage and adoption

- business priorities and requirements

Intranet teams also have the opportunity to conduct before-and-after measurements, allowing the benefits of the redesign to be clearly articulated.

Most of all, by building an in-depth understanding of staff and their activities, intranet teams can be confident that the new site will be successful and valuable.

Chapter 5

Define the intranet brand

What is the intranet for? An easy question to ask, but surprisingly hard for many intranet teams to answer. For the intranet design project to succeed, there must be a clear definition of the overall purpose and character of the intranet.

This goes beyond short-term redesign priorities such as 'making the intranet easier to use', 'fixing the navigation' or 'making the intranet work better'. It also says more than high-level intranet goals such as 'provide a one-stop shop for staff', 'provide all staff with the information they need to do their jobs', or 'provide a gateway to corporate information and tools'.

A clear definition of the intranet's purpose helps the project team to decide:

- what new capabilities and content to deliver on the site

- what features won't be included

- where to spend the project's limited time and budget

- how to determine success

- where the intranet should be going in the longer term

Ideally, the objectives for the design or redesign project would be derived from the overall goals for the intranet. Here again, teams may struggle. Thankfully there is an easy way to determine the overall nature and purpose of the intranet, at least at a high level.

What intranet will you be delivering?

There are many directions that an intranet can take, and many different aspects that can be focused on. You might, for example, want to deliver a:

Corporate intranet

This is the serious intranet, the 'men in suits'. It provides core corporate information and tools, such as HR, finance and IT. The intranet is used to disseminate company-wide messages, and updates from the CEO.

There is no social or fun side to the intranet, it's all about work.

Useful intranet

It's not pretty, but it is very useful. The focus is on saving staff time, streamlining business tasks, and reducing overheads. This is as much about the tools the intranet provides as it is about content.

The most useful intranet is the one used every day by staff, a key part of their standard working practices.

Rockstar intranet

The intranet is at the leading edge of innovation within the organisation, and is helping to drive cultural change and new working practices. The site is feature-rich, impressive, and engaged directly with staff.

More than just a corporate tool, the intranet is fun, friendly and collaborative. It's a key part of the company culture and DNA.

Or perhaps some combination of these. It is difficult to deliver all of these objectives in the one project. The more the intranet attempts to be 'all things to all people', the harder it is to manage and communicate. Within limited resources, design projects must also focus their efforts, to deliver a few things well, rather than many things poorly.

Define the intranet brand

Once an understanding has been gained of staff needs and issues, and overall organisational culture (Chapter 4), it becomes possible to define the 'intranet brand'.

The word 'brand' can mean many things. At a basic level, it is widely recognised as being associated with:

- logos
- colours
- naming
- identity

We will use brand much more broadly, focusing in addition on:

- purpose
- role
- character
- personality

By defining the intranet brand, we define the fundamental nature of the site. What does it set out to do? How does it aim to help staff and the organisation? How does it present itself? What does it deliver? And what doesn't it?

These are all important questions that can be answered with a clear definition of the intranet brand. It is important to highlight, however, that there is no single 'correct' answer to this.

Intranets prosper when they are direct reflections of the organisations they serve. While intranets can aim to progress the corporate culture by perhaps half a step, they must, on the whole, match where staff are at today.

Intranets can also play many roles, depending on what else is in place, and how staff work. On a factory floor, communication may be face-to-face, and news pinned on a noticeboard. The intranet plays a role as a source of procedural information and a way of accessing key operational tools.

In a global firm, the top-level intranet site may focus almost exclusively on corporate content and overall company vision. Country-level sites then play a more direct role in supporting daily work by staff.

What will be your intranet's brand?

Take a step-by-step approach

There are many ways of determining the nature and purpose of the intranet, but there is a simple approach that can be used quickly and effectively.

It involves the following steps:

1. Download the 'Microsoft product reaction cards', and make up a set of physical cards (more on this shortly).

2. Bring together key intranet stakeholders, and run a facilitated session where a handful of cards are chosen by the group to represent the intranet.

3. Supplement this with broader strategic input, practical considerations and staff input.

4. Flesh out the words into an overall description of the intranet.

By the end of this process, the intranet team will have a high-level description of the intranet brand, which can be used to shape the design of the site, as well as guiding the selection of content and functionality.

Microsoft Product Reaction Cards

The Microsoft Product Reaction Cards are a generic set of descriptive words that can apply to almost any product or service, whether electronic or physical. Created by the user experience team at Microsoft, they are freely downloadable from the following address:

www.microsoft.com/usability/UEPostings/ProductReactionCards.doc

The set includes words such as 'innovative', 'engaging', 'collaborative', 'effective' and 'useful'. These can be used to help define the future direction and role of the site, narrowing down the many possibilities to one clear direction.

The cards also contain a range of negative words, such as 'dated', 'overwhelming', 'irrelevant', 'busy', and 'complex'. These may be chosen by staff when asked to describe the current intranet.

(The cards have many uses beyond a visioning exercise, such as during user research or as part of discussions with stakeholders. Keep them close to hand, as they are always useful for filling a gap in activities.)

Print the words onto labels, and stick them onto filing cards, as shown in Figure 5-1. This produces a very 'hands-on' tool that works well in group settings as well as one-on-one discussions, and fits neatly in a briefcase or backpack.

Figure 5-1: The Microsoft Product Reaction Cards, transcribed onto filing cards.

Run the session

Bring together a group of key intranet stakeholders, along with anyone else who has valuable input on the design or delivery of the site. Running a facilitated session with this kind of group can be valuable in many ways, as it helps to create a common vision and direction.

Use the reaction cards as the basis of an activity to pick words that describe the desired intranet. Allow the group to select only eight in total, as shown in Figure 5-2. This number is carefully chosen, as 10 or 12 cards would be too easy, eliminating valuable discussion and debate.

In this exercise, the negative cards will be quickly discarded, greatly reducing what could otherwise be a daunting pile Many of the cards are synonyms, promoting discussion on which word is the best fit.

The outcome is a set of words that describe the intranet. In the process of narrowing down to just eight cards, implicit decisions have been made about the nature of the intranet.

The intranet may be mostly about content, leading to the choice of words such as 'accurate', 'comprehensive', 'trustworthy', 'reliable'. Perhaps collaboration

Figure 5-2: Eight words to describe an intranet. Just one of the possible outcomes from a session using the Microsoft Product Reaction Cards.

and enterprise 2.0 is the focus, with words such as 'personal', 'customisable', 'engaging'.

The focus may be on supporting business processes, with 'essential', 'efficient' and 'time-saving' being chosen. Words such as 'fun', 'meaningful' and 'empowering' may reflect a focus on organisational culture.

Regardless of the outcome, this simple group activity is a valuable first step towards defining the role of the intranet.

Refine the brand

The intranet should not be designed on the basis of the input of just a single group of stakeholders. The outcomes of the facilitated session should therefore be supplemented by other sources of input:

- other stakeholders throughout the organisation

- intranet authors and publishers

- findings from the needs analysis conducted with staff (Chapter 4)

- corporate goals and vision

- intranet best practices

Use these to refine and tune the outcomes of the group session. This may involve changing some words, or adding additional words. For example, the longer-term intranet strategy may call for a focus on task completion, suggesting that 'time-saving' should not have been discarded from the final eight cards.

Put flesh on the bones

By themselves, individual words have many meanings. Words that were read a certain way by the initial group of stakeholders may be interpreted very differently by the broader organisation.

The eight chosen cards therefore need to be fleshed out into a more complete definition. This can be done in many ways, but the simplest is to put some additional definition around each word.

An example is shown in Figure 5-3, adapted from a real intranet project at a mid-size government agency. This is a very succinct and effective way of stating where the intranet will be focused.

Example: vision for the intranet

To launch and maintain a relevant, useful and easy-to-use intranet that is engaging, innovative and sustainable. The intranet will connect staff, empower collaboration and be kept relevant by continuously planning and implementing six-monthly releases.

- Relevant. The intranet helps me in my daily work. It provides what I need, when I need it, matched to my specific requirements.

- Useful. It helps me get my job done, saving time and frustration. It's a business tool and interactive; not just a home for documents and news.

- Easy-to-use. I understand how to use it, and can quickly find what I need by browsing or searching. I didn't need much instruction, whether using the site or contributing to it.

- Engaging. I like using the intranet, and it's attractive and fresh. I have positive experiences every time I use the site, and I recommend it to other staff.

- Innovative. Our intranet is different from the ones I've used in previous jobs. Not just innovative for innovation's sake, it keeps improving and finding new ways of helping me in my work.

- Sustainable. The intranet project has been set up from the outset with a clearly defined scope, an ongoing management plan, and sufficient resources to make it a long-term success.

- Empowering. The intranet allows me to work better with my team, contributing ideas and interacting with others. It helps me achieve my goals and add value to the organisation as a whole.

- Connected. It keeps me in touch with my colleagues throughout the organisation, as well as with my other team members. I know what is happening, and can keep on top of key activities and changes.

- Collaborative. A lot of my work is now done on the intranet, instead of via email and on the server. Working with my team, I'm more productive and our activities are better managed.

Figure 5-3: A fleshed-out version of an intranet brand, providing supporting descriptions for each of the chosen words.

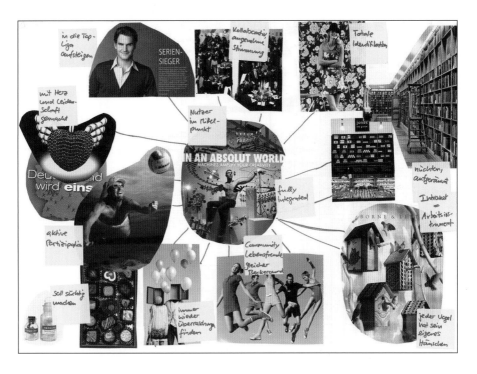

Figure 5-4: The intranet brand can be captured as a collage, with pictures chosen to represent intranet characteristics. (This collage is in German.) Image courtesy of Stimmt.

Consider a rich-media intranet brand

A variety of creative approaches can also be taken to capturing and communicating the intranet brand. For example, Figure 5-4 shows a collage (in German) created by intranet stakeholders. Pictures out of a magazine are used to represent elements of the intranet, and how it will work.

In this case, Roger Federer is labelled with 'play in the top league', drugs in the bottom left corner with 'intranet should be addictive' and the figure in the centre of the collage with 'fully integrated' and 'user in the centre of attention'.

This kind of activity can be fun for stakeholders, while still producing a powerful representation of the desired intranet. (Collages are often used by marketing agencies when defining website and product brands.)

Don't be afraid to experiment with any technique that will connect with staff and stakeholders. After all, hard work can still be fun!

Defining a broader strategy

This chapter has outlined a step-by-step approach that produces a high level statement of the intranet brand. This process has been designed to be straightforward and quick, allowing project teams to define key aspects of the intranet before starting into the design process.

This is, however, just one element of a solid intranet strategy. Derived from overall organisational objectives and priorities, there should be a clear understanding of the benefits that the intranet will deliver.

Chapter 6 will explore the role of intranet strategy, and will determine whether one is required as part of the current design or redesign project.

Outcomes

By the end of the process outlined in this chapter you will have a clear definition of the intranet's brand: what it will do, what it won't do, and how it will be presented.

This is the foundation for many decisions throughout the design process, including:

- selecting what functionality and content to include in the launched (or relaunched) intranet

- designing the visual identity ('look-and-feel') for the site

- choosing a name for the intranet

- determining what should go on the homepage

- resisting requests for non-core additions to the site

- maintaining a robust scope and plan for the intranet

Repeatedly return to the intranet brand, and call on it whenever there is debate amongst the team or stakeholders on what direction to go. The more the intranet brand is used to inform decisions, the stronger and more coherent will the resulting intranet will be.

Have you used these techniques to define your intranet brand? If so, we'd like to hear from you. We are always looking for additional examples to include in future editions of books and reports, so drop us a line at the following address:

contact@steptwo.com.au

Chapter 6

Determine your strategy, scope and support

Having spent time with staff (Chapter 4), there is now a clear sense of where the intranet can help (and where it can't). A deeper understanding of how staff work is also valuable throughout the design process.

Working with key intranet stakeholders, the 'intranet brand' now defines the character and focus for the intranet (Chapter 5). This gives the intranet clear priorities in terms of the functionality and content that should be delivered.

The temptation is to now jump straight into the design or redesign project, but pause a moment first. Like every other project, success rests heavily on having a good plan at the outset.

At a minimum, there must be a clear *scope* for the project. This locks in exactly what will be delivered on the new site, informed by what can be done within project constraints. This provides a foundation for project planning, including timelines and resources.

This is also the opportunity to go beyond the simple intranet brand, and to develop a more comprehensive *intranet strategy*. Taking a longer term and more strategic view, this positions the intranet in the context of business strategy and corporate priorities.

Above all, the intranet team must muster the necessary *support* from across the organisation to deliver the new site. Intranet projects are disruptive for many staff, and require considerable change management. They also need support from senior management and key stakeholders if they are to proceed smoothly and confidently.

Understand the organisational context

Intranets are the glue that connect organisations. They are used by all staff, and every business area has a stake. Intranets provide an entry point into corporate systems, and a communications channel that reaches every corner of organisations.

This is the power of intranets, but it is also the challenge. Intranet projects can often find themselves at the nexus of many different (and sometimes competing) priorities, and are dependent on many other business areas and projects.

To be successful, intranet teams need to build a good understanding of these organisational aspects at the outset of the design process. This allows decisions to be made, with confidence that they are realistic and achievable. It ensures that issues are identified early in the project, and plans in place to manage or mitigate them.

There are many things that intranet teams need to consider:

- *Senior management.* Who are the key members of the senior leadership team in relation to the intranet? What is their level of interest in, and awareness of the intranet? What senior management support will be required to deliver a new intranet? Who will be the overall project sponsor?

- *Key stakeholders.* Which are the business areas that have the greatest stake in the intranet? What groups are currently managing or publishing content? What roles will key groups play in the project? What is the overall governance model for the intranet?

- *Budget and business case.* Roughly what budget and other resources will be required for the project? Are these currently available to the team? If not, where will they be obtained? Will a business case or equivalent be required? When does the money need to be spent?

- *Decision making.* What approval or sign-off is required to start the project? What is the process for gaining this? How long will it take? What committees or groups are involved?

- *Timeframes and expectations.* Is there a specific deadline for the project to deliver a new intranet? If so, who specified this? What are the expectations of senior management and stakeholders of what will be delivered?

- *Project management methodology.* What methodology or approach is typically taken for these types of projects? What are the steps involved? What paperwork or other formalities are required?

- *Technology platform and IT support.* What platform will be used as the foundation for the new intranet? What is required to deploy it, and when will it be ready? How much will it cost, and who will pay for it? How long will the technology platform be in place? What level of support can be expected from IT? What resources will IT devote to the project? Are there technology issues that might impact on the project?

- *Other projects.* Are there other major projects that are happening, or due to happen, in parallel with the intranet? Is the intranet dependent on any other projects? Are there strategic or enterprise-wide initiatives that need to be considered when planning and designing the intranet?

- *Organisational culture.* What is the prevailing culture of the organisation? Is there likely to be resistance to change, and from where? What are the business and cultural issues that drive decision making? Are there any pockets of the organisation that have specific cultural considerations?

- *Intranet and project team.* Who will be working on the intranet project? What skills do they have? Are there any gaps? How much time will they have to work on the project? Who will be the project manager? Will additional resources be required, and where will they come from? What process is involved in gaining external assistance (consultants or contractors)? What impact will the project have on day-to-day management of the site?

Don't be daunted by this list! The important thing is to have answers for the key questions that will shape the intranet project, without getting trapped in 'analysis paralysis' before work even starts.

Intranet projects can range from a simple upgrade managed part-time by a single person, through to million-dollar projects based around an enterprise-wide technology deployment.

Larger projects naturally require greater support, planning and management. This may include a full business case, substantial budget, formal project management, and the direct involvement of a range of stakeholders. In contrast, smaller projects may be run more informally.

In all cases, intranet teams must have a sound understanding of the *mechanics* of the project, and 'how things are done' in the organisation. Even in the smallest of projects, stakeholder, technology and cultural issues will play a key role.

It is surprising how many redesign projects arise from a recognition of the failings of the existing site, and then proceed far along the process without having a clear idea of how the new site will actually be delivered. As they say: 'failing to plan' is the same as 'planning to fail'.

Is a strategy or roadmap required?

This book outlines a practical methodology for designing or redesigning an intranet that delivers a site that works from a usability and functionality perspective. Taking this tactical approach, insights from staff and the overall intranet brand are sufficient to guide the design process.

Intranets can, and ultimately should, play a strategic role within the organisations they serve. More than just a repository for content and a channel for news, an intranet should be a key business tool that underpins day-to-day work by staff.

To achieve this, the intranet team needs to go beyond the 'mini strategy' implied by the intranet brand. In most cases, a full *intranet strategy* will be needed that outlines a longer-term vision for the intranet. A concrete *roadmap* for the next 1–2 years then provides a phased approach to delivering this vision.

If a strategy and roadmap is required, this is the time to create it, before starting the design process. Creating a strategy, however, is not always necessary or appropriate.

Consider creating a strategy and roadmap when:

- *A mature intranet is being taken to the next level.* If the basics have already been addressed, a strategy can define the next steps for the intranet, moving it closer to the vision of a universal business tool. More experienced intranet teams also gain greater benefit from having a strategy.

- *A large, comparatively costly project is planned.* Smaller, more incremental projects often don't need to be supported by a top-level strategy. As the size and impact of the project grows, however, a strategy may be required to gain the necessary resources and support.

- *It is expected by senior management.* In some organisations, strategy documents are both required and routine. In these cases, the intranet should meet senior management expectations, in order to gain top-down support for the project.

Intranet teams can often find creating a strategy to be challenging and difficult. Depending on the current maturity of the intranet, it can be hard to articulate concrete business benefits and outcomes. The organisation may also lack strong corporate strategies that the intranet can be aligned with.

Creating a strategy also takes considerable effort and time, and will delay the start of the redesign. While a strategy and roadmap will ultimately be needed for every intranet, assess whether this is the right time, or whether the intranet (and the organisation as a whole) would benefit from a more tactical project.

Avoid creating a strategy and roadmap for the intranet when:

- *A brand-new intranet is being created.* If the organisation has no experience with intranets, it will be difficult to articulate a strategy and concrete benefits. Consider taking a more incremental approach, quickly launching a new site and then progressively improving it.

- *The project is smaller or incremental.* Many intranet projects focus on improving a few key aspects of the site, or address site issues via a series of ongoing improvements. In these cases, it is beneficial to have a vision of the 'end state', but a formal strategy and roadmap may not be required.

- *The existing intranet is immature and weak.* Many redesign projects are triggered by pain of a truly broken existing intranet. Not only will the project be forced to focus on fixing the basics, but the weaknesses of the existing site make it hard to articulate a business-focused intranet strategy.

- *It is hard to align the intranet with business needs.* Compelling intranet strategies demonstrate concrete business benefits and outcomes. This may currently be challenging for the intranet (and the intranet team).

- *Senior management haven't asked for one.* While a good strategy delivers many benefits, there are risks in delivering a weak strategy. If senior management is content to support the intranet without mandating a full strategy, consider starting straight into the redesign.

Engage with senior management and key stakeholders at the outset of the project to identify the need for an intranet strategy and roadmap. One-on-one discussions with key players will help the intranet team to understand motivations and expectations, as well as the political landscape.

Facilitated workshops can help to uncover organisational ambitions for the intranet, and the steps necessary to deliver it. The project sponsor should also be able to provide valuable advice to the team regarding the project approach.

It may strike some as strange that an intranet strategy is not considered a mandatory requirement for every intranet project. Intranet teams should always have a clear idea of where they are trying to get to. The greater the ambition and resources, the more a strategy and roadmap will be required.

The reality is that most intranets don't have a formal strategy, let alone a roadmap. It shouldn't be this way, but the absence of having these things should not stand in the way of making much-needed improvements to intranets.

This book therefore takes a pragmatic approach, matched to the current state of most intranets (and intranet teams).

Create a strategy (if required)

As discussed on page 19, this book focuses tightly on the design activities needed to deliver an intranet that works well for staff. Intranet strategy, and the development of a longer-term roadmap, is therefore out of scope for this volume. (It will be covered extensively in a future book just on intranet strategy and management.)

A strategy can be a small as a single sheet of paper outlining key objectives and deliverables. At the other end of the extreme, the strategy can grow to encompass enterprise-wide information management and knowledge management issues, painting a future vision that is three to five years away.

The most effective intranet strategies are closely aligned to business issues and priorities. Start with an understanding of corporate vision, mission, goals and plans. These are often captured in corporate documents and presentations (hopefully already published on the intranet!).

Don't be afraid to approach senior management, including the CEO, directly to discuss organisational priorities. While the intranet may not be on their radar, or in their thinking, they should be able to articulate business priorities. These discussions will also help the intranet team to strengthen their relationships with senior management, and to understand key managers' individual motivations and interests.

Once the business priorities have been identified, use these as a 'lens' to examine the findings from the staff research (Chapter 4). For example, a hundred different issues and ideas might have surfaced from across the organisation. By seeking ideas that also match business priorities, this longer list can be filtered down to a more targeted set of activities.

Taking this combined 'top-down' and 'bottom-up' approach means that the strategy and roadmap focus on activities that meet staff needs as well as organisational objectives.

Always match the strategy to the maturity and readiness of the organisation. While longer-term objectives will always aim to move the organisation forward in its operations and handling of information, any improvements delivered now must be adopted and used.

A good intranet strategy finds the balance between the future and the present, highlighting where the intranet can help in the bigger picture, while providing a roadmap consisting of practical and useful improvements.

Don't be afraid to go beyond a typical 'strategy document', to explore ways of targeting emotion and points of pain in a compelling way. A strategy is also a living thing, as much a process of ongoing engagement with the business as it is a written document.

Determine project scope

Even if a high-level intranet strategy and roadmap isn't produced, the intranet team must determine a clear *project scope* at the outset of the redesign. This must give a simple answer to a simple question: what will be delivered on the new site?

This includes:

- new or improved content

- additional functionality

- design changes

- underlying technology improvements

Start with a robust and pragmatic understanding of the organisational situation, as discussed on page 64. In practice, project constraints determine what can be delivered much more than goals and ambitions. The culture and working practices of the organisation will also determine how fast the project can proceed.

Make a realistic assessment of the resources that the intranet team will have its disposal, including staff time. If team members will be simultaneously designing a new site while maintaining the current intranet, reduce available time accordingly.

Assess potential site improvements, drawing on the insights gained from staff (Chapter 4) and business priorities. Use this to create a long list of new features or improvements.

Working with the project sponsor and key stakeholders (including IT), identify a scope that is:

- *Achievable within project constraints.* Above all, the planned improvements must be possible to deliver by the nominated deadline, with the resources currently available, and within project constraints.

- *Useful for staff.* The new intranet must be tangibly and visibly better for staff, addressing common complaints and focusing on important tasks.

- *Valuable for the organisation.* The features and content delivered by the intranet should provide concrete benefits for the organisation, and be aligned with corporate priorities and goals.

- *Helpful in building momentum for further improvements.* Each intranet project should build support for future changes and enhancements.

There is no 'right' scope, as shown in Table 6-1. Projects can range from incremental improvements through to wholesale redevelopments, with many other permutations and variations. If in doubt, err on the side of a smaller scope. If it isn't clear that a feature or improvement can be delivered, particularly for reasons of technology uncertainty, drop it from the project.

Fundamentally, it is better that the team goes into the project with a realistic scope that can actually be delivered, rather than 'stretch targets' based on optimistic assumptions. In the unlikely situation that everything is completed early, additional features can be delivered. The alternative is often to drop hoped-for (and promised) features when time becomes tight.

Document the scope in a way that matches the size and complexity of the project. This can be as simple as a list of bullet points on a sheet of paper, or as complex as a full project plan following a formal project management methodology. Done well, this information can also be used as the basis for project communications, as shown in Figure 6-1.

Once the project scope is finalised, have it signed off, and locked in. Following good project management principles, protect the project against scope creep, and focus ruthlessly on delivering planned improvements. (If some form of 'agile' methodology is used in the organisation, this can provide a more nimble alternative.)

The process of determining a scope need not be complex, although it is remarkable how many teams progress well into a redesign without locking down what exactly will be delivered. Other teams fall into the trap of planning optimistically, rather than pessimistically, and end up promising improvements they can't deliver.

Resist any temptation or pressure to jump straight into the design process, and spend the time required to develop a clear scope and plan for the project. This will save heartache later in the project, and will help to ensure that what is promised is actually delivered.

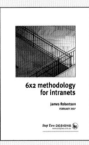

6x2 methodology for intranets
by James Robertson

Based on experience across many different organisations, the '6x2 methodology' provides a simple and practical way for intranet teams to determine project scope. Balancing goals and constraints, it identifies improvements that are both doable and worth doing.

For more information:
www.steptwo.com.au/products/6x2

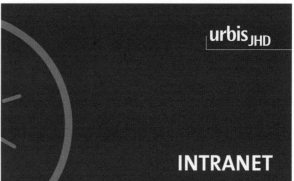

KM

INTRANET

The UrbisJHD intranet exists to make life easier for staff.

It provides useful information and tools that help you work and network efficiently and effectively.

GOALS

The intranet will:
- save you time
- help you connect to the right people
- help staff understand the whole firm
- evolve as UrbisJHD changes
- be consistently accurate and useful
- be a tool that you trust and use

PRINCIPLES

In order to achieve the above goals we will:
- publish content and tools that clearly add value
- often engage with staff to ensure that content and tools do add value
- keep the site up to date and remove old context
- make using it as easy as possible

BUILDING MOMENTUM

Until recently, the most used areas on our intranet were the phone list and social event photos. Now we've got:
- a snazzy new look & feel with an informative home page including "what's new"
- an improved phone book with photos and 'show desk' maps
- new organisational charts
- news articles featuring UrbisJHD
- new library page with acquisitions lists and featured books
  ... and many more useful resources

NEXT STEPS

Next, we'll deliver:
- a new marketing portal full of useful tips and tools
- a search facility to help you find things on the new intranet
- a new library catalogue so you can more easily find the resources you need

AFTER THAT

Later this year:
- actively updated topical pages, such as residential development and research
- we'll launch DIG - the new Digital Intranet GIS - that provides information and maps at your fingertips
- provide new pages that allow you to browse our electronic information resources

GET INTO IT!   **http://web01/urbisjhdintranet**

Figure 6-1: Intranet marketing and communications materials can be quickly developed from the project scope. Example courtesy of Urbis.

Four possible intranet projects

Project	In scope	Out of scope
Incremental improvement (Fix key issues with current site)	Redesigned homepage Update to basic page templates Restructure of one or two key site sections Photos in the staff directory More online forms Tweaks to search engine Further support for content authors	Change of underlying technology platform Overall site restructure and redesign Site-wide content review Collaboration or social tools Major new functionality or applications Broader knowledge management or information management objectives Anything that requires significant budget, resources or IT support
Core intranet redesign (Improve site structure, design and navigation)	Redesigned homepage Overall restructure of intranet, including new navigation and top-level categories Redesign of basic page templates Review and content cleanup New publishing processes and policies Review of intranet governance and management Improvements to intranet search Enhanced staff directory Additional collaboration capabilities. A few key pieces of intranet functionality	Change to the underlying technology platform New search engine Major new social tools Extensive application development or IT involvement Broader knowledge management or information management objectives

Table 6-1: Intranet projects can range from small-scale incremental improvements to wholesale redevelopments. (These are just a few examples of possible approaches.)

Four possible intranet projects

Project	In scope	Out of scope
Major intranet redevelopment (Redesign on new technology platform)	Development of intranet business case and metrics Overall redesign and restructure of the intranet Selection and implementation of new publishing platform New search engine Complete content review, cleanup and migration to new platform New publishing processes and intranet governance Deployment of collaboration environment Integration of social features into core intranet Integration with records management system	Substantial application development Extensive customisation of intranet platform Intranet personalisation or targeting Enterprise search Enterprise-wide information management
Intranet as a business tool (Improve processes and functionality)	'Single sign-on' across intranet-based applications New online applications to address key business needs Streamline selected business processes Replacement of PDF-based forms with online equivalents Tailored homepage for staff that provides day-to-day tools alongside corporate information Establish ongoing process for further business tools	Redesign or restructure of the existing intranet Content review and cleanup Replacement of existing publishing tools or search engine

Table 6-1: Intranet projects can range from small-scale incremental improvements to wholesale redevelopments. (These are just a few examples of possible approaches.)

Implement more than just a restructure

The focus of the redesign efforts to this point has on solving the fundamental complaint from staff: 'I can't find anything!'. A new structure has been developed and tested with staff to ensure that it is easy and intuitive. Page designs have been created that balance stakeholder and staff needs, while meeting core usability principles.

This is worthy and valuable work, but it's not enough. Where the team is redesigning an existing intranet, this work is all focused on solving yesterday's problems, the legacy issues with the site. Even when creating a new intranet from scratch, good navigation and designs are just a foundation for a great site.

When successful, staff take intranets for granted. If they work as expected, at the time when they are needed, staff are satisfied. If they can't find what they need, they are frustrated.

Not knowing the huge amount of work that has gone into the new intranet, staff are likely to ask: what else are you giving me?

For this reason, intranet projects should deliver more than just a redesign. In addition to improving usability, additional functionality or solutions should be delivered. These might be small (but important), or larger-scale deliverables. Possibilities include:

- new staff directory functionality

- an 'office locator' tool

- collaborative capabilities

- replacement of key PDF-based forms with actual online forms

- room booking system

- employee self-service

- knowledge base for the call centre

- intranet functionality on mobile devices

These are just a small selection of the possible improvements, with some addressing core intranet functionality, while others tackle business needs. Careful consideration should be given to the functionality that is launched.

More than just a gimmick, it should address the 'what's in it for me' factor for staff, as well as delivering concrete business benefits. Promote these enhancements alongside the redesign of the intranet, potentially giving them priority over behind-the-scenes improvements to the underlying intranet.

Build the case for change

The current intranet is broken. Alternatively, there is no existing intranet, and one is needed. Even when this is recognised across the organisation, this doesn't automatically translate into a commitment of the resources needed to make it better

One of the most common mistakes made by intranet teams is to push forward on the project without gaining the necessary support. Failing to recognise the scale of support needed, some teams are brought to a halt by unmanaged stakeholder and staff issues.

Intranet projects are disruptive for many, and will impact productivity and familiarity with the site. Migrating to the new intranet, and its supporting ways of working, will also require large amounts of effort and time. An intranet redesigns may be just one of a number of major projects across an organisation, competing for time, money and attention.

The starting point for building support is to create a sense of urgency. This involves building a strong case why the current situation cannot continue, and why change is necessary.

This is particularly important when garnering support from senior management, who will need to back up the intranet team throughout the project. It is not enough to simply highlight current usability and design problems; intranet teams must justify the business reasons for committing resources to the redesign project.

There are many ways of building the case for change. Uncovering examples and stories from staff (Chapter 4) can often produce the most powerful results, particularly when supported by quotes and videos. Business risks can also be a strong driver for many organisations. In all cases, seek ways of aligning the intranet project with business objectives and goals, beyond just addressing intranet design and management deficiencies.

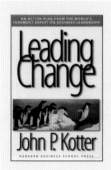

Leading Change
by John P. Kotter

This is the definitive guide for managing projects that have a large impact on organisations and their cultures. It provides a step-by-step approach that can be applied to any project, and it's directly useful for intranet redesigns.

Also consider obtaining the companion volume: "The Heart of Change".

ISBN: 978-0875847474

Gather support

Redesigning a well-established intranet involves major cultural change, as does creating a brand new intranet where one hasn't existed before. For anything beyond the smallest of tweaks or incremental changes, intranet projects will have an impact on business areas, content owners, authors, and end users (staff).

A major redesign project will touch every aspect of an intranet, including:

- overall site structure and design

- content ownership and management

- authoring and publishing processes

- resources and responsibilities

- technology and infrastructure

- day-to-day use by staff

Intranet teams must spend enough time and effort building support for the project ahead at the outset of any major redesign.

Table 6-2 lists some of the key players in intranet projects. These range from the senior leadership team down to intranet authors and publishers. All have a role to play in intranet projects, and they must provide support from day one through to post-launch.

When building support, ensure there is a clear message that encompasses:

- current problems and reasons for change

- benefits that will be delivered

- what will change and when

- why the specific changes are necessary

- what will be required in terms of support, money, resources and time

- how working practices will be different

Use a mix of approaches for building support, including one-on-one meetings, group discussions, strategy sessions and formal documents. Build an understanding of the interests and motivations of each group, and position the project in a way that reflects their needs.

Above all, don't proceed on the design or redesign project until there is sufficient support to ensure it continues smoothly and successfully.

Group	Role	Support required
Senior management	Organisational strategy Business leadership	Endorsement of intranet project Visible support and backing for the project
Project sponsor	Overall 'owner' of the intranet	Leadership of intranet project Project budget Allocation of project resources Assistance with internal politics
Management committees	Decision making and approval for major projects	Endorsement of intranet project Approval of project scope and outcomes Alignment with other initiatives Project budget
Major stakeholders	Owners of key systems (eg IT, finance) Owners of key processes (eg communications, HR)	Engagement with intranet project Resources to support project Support for making necessary changes
Business areas	Owners and managers of intranet content Representatives of major staff groups (end users)	Commitment of resources within their areas to migrate and improve content Support for intranet deployment
Authors and publishers	Creating and updating intranet content	Understanding of new intranet design and governance Time and resources to migrate and update content
Staff	End users of the intranet	Understanding and acceptance of changes Uptake and usage of new site and site features

Table 6-2: There are many stakeholders across organisations who need to support and assist intranet projects.

Outcomes

A concrete scope has been determined for the intranet project, defining what will be delivered when. Planned improvements will help staff in their work, as well as benefiting the business. They are also achievable, within project constraints, organisational culture and working practices.

If needed, an overarching intranet strategy has been developed which outlines the longer-term objectives for the site. This includes a concrete roadmap of improvements over the coming few years.

Most importantly, time has been spent building support with key stakeholders and business areas across the organisation. There is a clear and common understanding of why the intranet must be improved, and what will be involved in delivering a new site.

The intranet team is now ready to start the design process!

Chapter 7

Intranet design methodology

As outlined in Chapter 2, staff come to an intranet at the point of need, to find a piece of information or to complete a task. Intranets must be well-designed to meet these needs quickly and easily.

There are many traps to fall into, as highlighted in Chapter 3. One of the biggest is producing a design that makes sense to the designer and business stakeholders, but not to the staff who are expected to use the site.

What is needed is a robust methodology that will deliver a new or redesigned intranet that works well. This methodology must fulfil several key criteria:

- achievable within typical resource and time constraints

- manages the sometimes conflicting opinions and ideas of all involved

- delivers a site that is intuitive and efficient for staff

- scales from the smallest to largest intranets

- provides confidence that a great intranet is being delivered

This chapter outlines a 'user-centred design' methodology that meets these criteria. Commonly used across both websites and intranets, this approach is robust and well-tested.

The fundamentals will be outlined in this initial chapter, and most of the remaining chapters in the book will explore each step in turn, providing practical details and insights.

Intranet design challenges

Before outlining the methodology, it is worth revising some of the key challenges confronting intranet design projects:

- *Very large site.* Intranets can grow to be thousands or even millions of pages, and redesigning such a site is no small matter.

- *Diverse staff needs.* Intranets serve the needs of all staff, but their daily tasks, key needs, and working environments vary greatly. This means there is not a single 'intranet user' audience that can be targeted.

- *Organisational factors.* Complex organisations, including large businesses, have many different areas, activities and processes. These impact on the design and management of the intranet.

- *Many opinions.* Every manager and staff member is a stakeholder of the intranet, and many will have opinions about what needs to be done, and how the intranet should be designed.

- *Competing priorities.* With so many business needs and opportunities across the organisation, the redesign project can be battered with requests (or demands) from senior managers and other stakeholders.

- *Global vs local.* Some information is company-wide, such as corporate news, but much of the day-to-day information needed by staff is local, specific to their role or business unit. These two needs must be carefully balanced when designing the intranet.

- *Technology considerations.* Every publishing tool used to manage intranets has strengths and weaknesses. Intranet projects need to work within the constraints of the tools, while making the most of features that are provided.

- *Tight timeframes.* While some intranet projects have the luxury of time, this is uncommon. With the delays involved in kicking off the project, there is often considerable time pressure to deliver a new site.

- *Limited budget and resources.* Intranet teams may be small (or nonexistent!), with tight budgets and staff constraints impacting every aspect of the redesign project.

While no methodology can make all these challenges go away, the approach outlined in the coming chapters will do much to manage or mitigate these issues.

User-centred design

The foundation for a successful intranet design or redesign project is to follow a 'user-centred design' (UCD) methodology. This provides a toolbox of concepts and techniques that involve staff in a structured way throughout the design process.

In practice, user-centred design is made up of two related professional fields, usability and information architecture, as shown in Figure 7-1. Together, these give intranet teams a well-tested approach for designing sites that work.

Usability outlines a set of principles and approaches for ensuring that a site is easy to use. The main technique is usability testing (Chapter 13), in which staff attempt real-life tasks on the intranet while being observed. The observers don't tell the usability test participant how to use the system and don't answer questions. It is as if the participant were doing the tasks alone.

The usability test identifies the key usability problems with a site, which enables them to be fixed. This can be done early in the project on prototypes, as well as later in the project on draft intranet sites.

Information architecture (IA) mostly focuses on the structure and navigation of the site, including how search works. This includes a particular focus on organising content, and identifying the right names for things.

Information architecture provides a number of practical techniques that can be used to get these aspects right. These include card sorting (Chapter 9), which helps the team understand how staff think about information, and tree testing (Chapter 11) which allows the structure of the site to be tested.

> For more on these two fields, read the introductory articles "What is usability?" and "What is information architecture?":
>
> www.steptwo.com.au/papers/kmc_whatisusability
> www.steptwo.com.au/papers/kmc_whatisinfoarch

Usability and information architecture are widely used around the globe. There are many books on these topics, as well as professional associations such as the Usability Professionals Association (*www.upassoc.org*) and the Information Architecture Institute (*www.iainstitute.org*).

Like any professional fields, there can be disagreements about the names of things, and where the boundaries lie between disciplines. For intranet and project teams, this doesn't really matter. In any real-life project, techniques from both usability and information architecture (and other supporting fields) will be used, so we will lump them all together to simplify discussions.

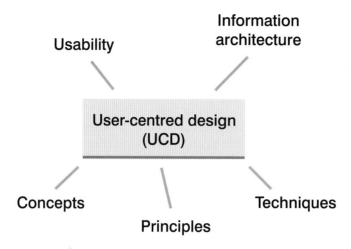

Figure 7-1: User-centred design (UCD) is an umbrella for the fields of usability and information architecture, which provide useful concepts, principles and techniques for the intranet designer.

It is not the goal of this book to provide a comprehensive textbook on the disciplines of usability and information architecture. As highlighted in the introduction, this book focuses on providing business teams with the information and tools they need to deliver successful intranets.

The coming chapters will introduce key techniques and concepts, going straight to the heart of what is needed day-to-day in an intranet design project. Throughout the sections, references will be made to further resources, such as the must-have book listed below.

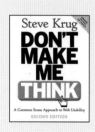

Don't Make Me Think
by Steve Krug

The single best book on usability for business people. Starting from key principles, it walks through how to design a site, and provides useful resources such as a script for usability testing sessions.

This is a must-have for every intranet team. Buy several, as you'll want to loan some out to content owners and stakeholders.

ISBN: 0321344758

Intranet design methodology

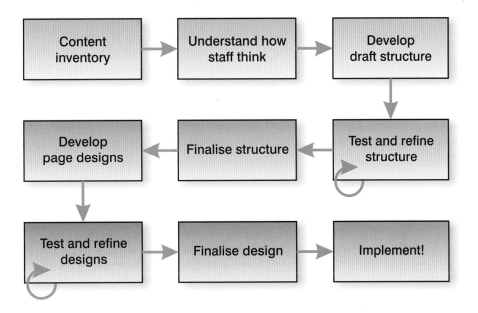

Figure 7-2: A user-centred design methodology for intranets.

Figure 7-2 outlines a core methodology for designing or redesigning an intranet. It involves the following steps:

1. Understand staff needs and business requirements by conducting user research with staff across the organisation (already covered in Chapter 4).

2. Build an understanding of how staff think about information, using a technique called card sorting (Chapter 9).

3. Develop a draft structure (information architecture) for the intranet based on the results of the card sorting, needs analysis, best practices, and other sources (Chapter 10).

4. Test and refine the draft structure, using a technique called tree testing to conduct hands-on testing of the draft structure with staff, and then revise accordingly (Chapter 11). This is an iterative process.

5. Develop draft designs including page layouts for key pages such as the homepage of the intranet (Chapters 12 and 14).

6. Test and refine the design of the intranet, conducting task-based usability testing with staff to ensure the designs are easy and intuitive, and revising where required (Chapter 13).

7. Finalise the designs, and document to the required level.

8. Implement the designs in the content management system or other intranet technology. Launch and celebrate!

As noted above, the coming chapters will explore each step of the methodology in turn, providing details and examples. By the end of this process, you will have created an intranet that works well for staff, delivers clear business benefits, and is on a path towards future improvements.

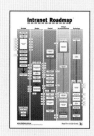

Intranet Roadmap

This outlines all the activities involved in designing or redesigning an intranet. Focusing on the sequencing of steps and overall project planning, this is an ideal companion resource to this book, and will help to align all team members and stakeholders.

For more information:
www.steptwo.com.au/products/roadmap

Design vs redesign

As discussed on page 20, there are two main types of intranet projects. In 'brownfield' projects, the task is to redesign an existing intranet. 'Greenfield' projects, by contrast, are creating a brand new intranet for the first time in an organisation.

The primary focus of the methodology outlined in the previous section (and the book as a whole) is the redesign of an existing intranet. This is the challenge confronting most intranet teams, as the majority of organisations have had an intranet in place for many years.

Some project teams are lucky enough to create a new intranet from a blank sheet of paper. These projects are uncommon, and have their own challenges. The user-centred design methodology also applies to these projects, and is crucial to ensuring the intranet starts out on the right foot.

Throughout the coming chapters, where there are special considerations for designing a new intranet, they will be highlighted. Greenfield projects can also benefit from understanding how redesigns progress, and therefore how to avoid the mistakes that created the need for a redesign in the first place.

Consider accelerated methodologies

There may be instances when it isn't possible to go through all of the steps outlined in Figure 7-2. These might include:

- creating a brand new intranet

- taking an iterative approach, rather than a once-off redesign

- insufficient budget or resources to conduct all activities

- timeframes that are too tight

- limited engagement and support from stakeholders and senior management

- site that is too small to justify the full methodology

Chapter 16 outlines a number of options that can be considered in these situations, which target limited time and resources to key activities and techniques.

Explore variations on a theme

What has been outlined is a solid, widely understood approach to user-centred design, and it is typical of the approach taken by firms across the globe. That being said, there are as many variations on design methodologies as there are experts and consultancies.

In most cases, these are just variations on a theme, different in one or two aspects, but otherwise very similar to the approach outlined here. In other cases there may be more fundamental differences, but the methods still draw on the basic principles of usability and information architecture.

These variations are generally not significant, and relate to the backgrounds of staff, or the personal experiences of those running design projects. (Many professionals have organically evolved a toolbox of techniques over time, based on experience and project work.)

This should not, therefore, be viewed as an iron-clad, inflexible methodology. There is plenty of scope to vary the methodology to fit individual circumstances, to mix-and-match particular techniques, or to take different approaches where needed.

When working with external experts or professionals, expect differences in terminology and approach, which shouldn't impact on the intranet project.

When finalising the methodology to be used, ensure it meets the following criteria:

- involves staff throughout the design process

- includes task-based testing with staff at key stages of the process

- avoids relying too much on 'golden rules'

- reflects both business priorities and staff needs

- provides reasons for design decisions that will be understood by stakeholders and content owners

- gives confidence that the right solution is being delivered

These criteria can also be used to assess the methodologies proposed by consultancies or outside experts (see below for more on this).

Gain professional assistance where needed

Good approaches to design are practical and achievable for any intranet or project team, regardless of their background or experience. Many of the techniques outlined in this book are not rocket science, and quickly provide useful insights and findings.

Every intranet project can take advantage of the benefits provided by a user-centred methodology, and much or all of the work can be done in-house.

There can, however, be benefits to gaining outside professional assistance when designing the intranet. While most steps in the methodology are easy to execute, there are a few stages that benefit from professional experience and expertise.

These are the two key design stages: creating an initial structure for the intranet (Chapter 10), and producing draft designs for key pages (Chapter 12). As will be covered in these chapters, there is no 'magic formula' for these steps. Instead, there is a mix of good design principles and creative thinking.

By bringing in an external specialist, you can draw upon the experience gained across many organisations. These can reduce the time taken to develop the designs, as well as producing more polished and robust designs.

Intranet projects are often short of time or staff, and bringing in additional support can help to address these gaps. This frees up the intranet team to focus on stakeholder management, organisational change and site management, while the new designs are being intensively developed.

A wide range of organisations can provide assistance to intranet teams, including usability consultancies, information architecture experts, web development agencies, and knowledge management firms. Look for a good mix of practical methodologies and in-depth intranet expertise.

Where appropriate, the value of bringing in external assistance will be highlighted. Never be afraid, however, to experiment or build skills in-house. The intranet redesign project is not a 'one-shot' exercise, and the site will continue to evolve for many years to come, giving further opportunities to address issues and deliver improvements.

Outcomes

The intranet team now has a concrete and practical methodology to follow when designing or redesigning the site. Involving staff throughout the design process ensures that the new intranet will be easy and productive to use.

Taking a structured approach will also help the intranet team to manage stakeholder expectations and opinions, and will build confidence that the right design has been delivered.

Chapter 8

Conduct a content inventory

When talking about long-standing intranets, teams often use metaphors such as jungles, forests, mazes or rabbit holes. Intranets grow organically over time, with individuals and business units publishing content to meet their needs, or the perceived needs of others.

The net result is a lot of content. A moderate-sized intranet is typically thousands of pages in size, larger intranets can easily be 10,000 or 100,000 pages. Particularly active organisations have intranets measured in millions of pages.

It is not practical for any one team or individual to keep up with all these changes and additions to the intranet. Despite that, the new intranet must find a home for all this content, at least the content that is worth migrating.

The starting point for a redesign is to conduct a *content inventory* (Figure 8-1). This creates a map of the existing content on the site, providing a key input into techniques such as card sorting (Chapter 9), as well as the creation of a new site structure (Chapter 10).

More than just a map, a content inventory provides a to-do list. Invariably, when the full breadth of pages is looked at, considerable amounts of dead or irrelevant content will surface. This can be cleaned up in parallel with the design process, ensuring only good content is migrated.

If you are creating a brand new intranet where one hasn't existed before, there obviously isn't an existing site to inventory. Skip this step, and proceed directly to Chapter 9.

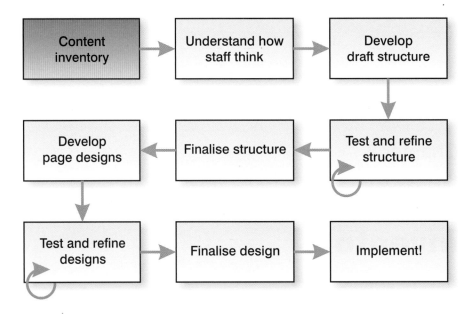

Figure 8-1: The result of a content inventory is a map of the existing site, as an input into the design and migration activities.

Conducting the inventory

A content inventory is an exhaustive examination of every page on the intranet. Start at the homepage of the site, and follow the links into each section. Work outwards, capturing details as you go.

The results of a content inventory are typically captured in a spreadsheet, as shown in Figure 8-2. This includes details such as:

- page name

- page location (URL)

- brief summary (where needed)

- content owner (may need to be filled in later)

- whether the page suffers from ROT (redundant, out-of-date or trivial)

By the end of the inventory, the intranet team should have a clear idea of what's on the site, including major groups of content, and key areas that need improving.

A	B	C	D	E	F
Page ID	Page Name	Link	Page Type	Owner	ROT?
1.0.0	For Staff	http://www.xyz.edu/staff/	paras with embedded links to webpages	Comms team	
1.1.0	Dean's Office	http://www.xyz.edu/staff/deans office/	Paragraphs (list of links)	Dean's office	
1.1.1	About the Dean	http://www.xyz.edu/staff/deans office/about	Text and picture	Dean's office	Redundant
1.2.0	External Relations	http://www.xyz.edu/staff/extern alrelations/	Paragraphs with links to pages	External relations	
1.2.1	2009 Event Schedule	http://www.xyz.edu/staff/extern alrelations/events/2009	List of items	External relations	Out of date
1.2.2	2010 Event Schedule	http://www.xyz.edu/staff/extern alrelations/events/2010	List of items	External relations	
1.3.0	Operations Portfolio	http://www.xyz.edu/staff/operat ions/	para - list of links to area pages	Comms team	
1.4.0	Research Portfolio	http://www.xyz.edu/staff/resear ch/	paras with links to EXTERNAL pages	Comms team	
1.5.0	Teaching and Learning	http://www.xyz.edu/staff/tl/	paras with list of links	Teaching & learning	
1.6.0	Committee meeting schedule	http://www.xyz.edu/staff/commi ttees/	table - onscreen and pdf (17kB)	Unknown	Out of date

Figure 8-2: The results of the content inventory are typically captured in a spreadsheet, noting key details for each page on the site. (Sample spreadsheet showing only some of the details that are captured.)

If a content management system or equivalent is in place, it may be possible to automate at least part of the content inventory. Many systems provide a report that gives a complete list of pages on the site, in a format that can be exported to Excel. Metadata such as creation and review dates can also give valuable insight into the currency of content.

Unfortunately, even when a useful report can be extracted from the content management system, it will still be necessary to examine the pages to determine whether they are useful or not.

For a very large site, the content inventory process may be difficult to complete in a single activity. In these situations, conduct a two-stage process.

Start by building up a high-level map of the site, as an input into card sorting sessions (Chapter 9) and the creation of a new site structure (Chapter 10). A more detailed inventory can then be conducted in parallel with the design process, in advance of the content migration (Chapter 15).

Jeffrey Veen from Adaptive Path has written a great overview of conducting a content inventory, including a very useful spreadsheet that can be downloaded and used:

www.adaptivepath.com/ideas/essays/archives/000040.php

Who should do the inventory?

Conducting a content inventory is far from exciting, particularly when the intranet is large, sprawling and out of date. At a minimum, it will take several days to complete, and could easily take several weeks.

There is naturally a temptation to delegate the content inventory to a junior team member, or to use temporary staff. University students might also seem an attractive option.

While they are able to fill in the details, it will be difficult for them to assess whether information is current or useful. (Temporary or external team members will know little about the organisation itself.)

Delegating in this way also means that the 'mental map' of the site is being built up in the head of someone outside of the core project team.

For these reasons, the content inventory needs to be conducted by the team members who will be developing the new site structure (Chapter 10). Where appropriate, 'sub inventories' can be conducted by owners of individual sections, before migrating content.

Ensure every page has an owner

As organisations restructure, and staff move on, intranets end up with content that is 'orphaned'. This will show up in the content inventory as pages (or whole sections!) with no owner.

Not actively owned or maintained, this content is often out of date, irrelevant or no longer needed. (If there isn't anyone responsible for updating it, how can the content ever be current?)

When it comes to the new intranet, every page must have an owner, without exception. This may be difficult to achieve in practice, as business areas may be reluctant to take on ownership without being given additional staff or resources. Some content may also be split in ownership between several different groups.

Despite these challenges, a simple rule must be applied when it comes to content migration (page 207): only migrate pages that have an owner. This can often be the hook for resolving content ownership.

If after talking to business units there is still a significant amount of content that is un-owned, publish this as a list, indicating that it will be deleted as part of the move to the new intranet. Whoever cries out 'that can't be deleted!' is the owner. Many teams have found this to be a simple but effective strategy for resolving content ownership.

Improve intranet content

Intranets are nothing without good content. This makes content an important part of any design or redesign project (although it is only peripherally touched upon in this book).

Good content is:

- useful

- accurate

- complete

- up-to-date

- trustworthy

- easy to read

- concise

- targeted to audience needs

- delivered in a suitable format

- cross-linked

Some of the content will be owned and managed by the central intranet team, but most will belong to individual business units. The content also varies greatly in importance, with core corporate content sitting alongside project updates and meeting minutes.

Establish suitable content standards and guidelines, taking a graduated approach that puts greater emphasis on the quality of critical content. This ensures that limited resources within central and decentralised areas are targeted to the highest value information.

Work with content owners and stakeholders to establish appropriate governance and management of the content. As part of an intranet design or redesign project, much of the heavy lifting will be done by business units updating their own content. The central team should play a supporting and mentoring role to assist this work.

There will often be value in also targeting key intranet areas for more intensive work. By allocating central resources, major sections such as HR or IT can be substantially improved during the project. This delivers clear benefits for staff, and ensures that these key areas deliver content to the highest possible standard.

Start cleaning early

Content migration is often the single biggest task in an intranet redesign, particularly when the intranet is very large. In some cases, it has taken more than 6–12 months to completely move content off the old site into the new designs and structures.

Over time, intranets accumulate an ever-increasing amount of out-of-date or dead content. Like junk stored in an attic, it is easy for it to accumulate until moving house forces a complete clean-up.

Many intranet teams have found that a thorough review of means between 50 and 90 per cent of pages can be deleted instead of migrated.

Getting rid of so much old content will dramatically shrink the content migration.

To ensure that the content migration doesn't hold the project up, start cleaning up content early. This can be started as soon as the content inventory has been completed, and suitable guidelines established for the new site (as discussed above).

Coordinate with business units and content owners to ensure that they understand the work that is required. Touch base with them throughout the design process to assess their progress, and to encourage further clean-ups.

Human nature being as it is, it will always be difficult to get everything cleaned up in advance of the migration itself. Like the junk in the attic, it's only when it comes to packing things into boxes that the hard decisions are really made.

Start early with content owners, but plan to do a big 'push' towards the end of the project once the new structures and designs have been finalised.

Outcomes

Having conducted a comprehensive review of intranet content, the intranet team now has a content inventory listing every section and page. This provides a 'mental map' of the whole site, and is a key into card sorting (Chapter 9), and the creation of new site structure (Chapter 10).

The content inventory has also uncovered a lot of ROT (redundant, out of date and trivial). Once content owners are identified for all content, they can be encouraged to clean up the intranet well in advance of the content migration.

The simple task of looking through the intranet can often identify thousands of pages that can be deleted immediately. That's a lot of pages that don't need to be restructured or migrated.

Chapter 9

Understand how staff think

Having made use of the needs analysis techniques outlined in Chapter 4, the intranet team will have a much clearer picture of daily working practices and unmet needs. If personas (page 41) have been created, these bring these insights into the design process in a very rich way.

It is still possible for intranet teams to fall into the trap of designing for themselves, creating a design that only makes sense to them as the designers. Teams need to be constantly vigilant against this, and always main a clear focus on delivering a site that works well for staff.

Beyond identifying staff needs, it's vital to get into their heads, to understand how they think about information, and how they look for content on the intranet. Teams can use this insight when developing a new site structure, ensuring that the new site isn't just a continuation of 'how it's always been done'.

At the outset of the design process (Figure 9-1), a technique called *card sorting* can do much to help. The most widely used technique from the field of information architecture, this provides a hands-on technique that can be used with a wide range of staff.

Within a group session, staff collect together intranet information in ways that makes sense to them, and in the process come up with ideas for top-level intranet navigation. This provides a valuable input into the design of the new intranet that can be supplemented by the results of the needs analysis, best practices, and professional experience. It is also valuable from a staff engagement perspective, building end-user support for the new designs.

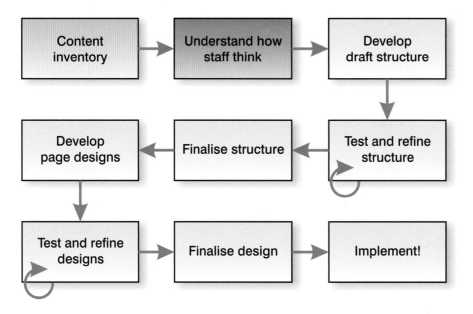

Figure 9-1: Card sorting sits at the beginning of the design process, providing insight into how staff think about information on the intranet.

Introducing card sorting

At its core, card sorting is a straightforward technique that can be quickly prepared and widely run:

1. Create a list of topics, and write these onto filing cards.

1. Bring together a group of staff together for a card sorting session.

2. Provide them with a pile of cards, each labelled with a piece of content or functionality that might be included on the intranet.

3. Ask the participants to group the cards into piles that make sense to them.

4. Once the piles have been determined, ask them to create a label for each pile.

5. Run several sessions with different groups of staff, and consolidate the results.

6. Use the groups and labels as an input into the site structure.

Create a list of topics

The first step in planning a card sorting session is to determine the list of topics that will be written on the cards.

Each topic represents a piece of content or functionality that will be delivered as part of the new or redesigned intranet.

If you a redesigning an existing intranet, there are many sources of information:

- results of needs analysis activities (Chapter 4)

- content inventory of the existing intranet (Chapter 8)

- future content and functionality requirements

Of these, the biggest source of topics is typically the content inventory of the whole site. In addition to assessing individual pages, this should uncover broad topic areas and key intranet functionality.

Topics can also be derived from offline documents and printed materials, and from descriptions of business groups and processes.

Use as many sources as possible. This helps to create structures that work for current content and functionality as well as for future materials. Adding new items in the future should therefore require minimal rework if the structure is designed to cater for these items.

Card sorting is a great 'do-it-yourself' technique. It is simple to set up and run, and the results are obtained almost immediately (although careful analysis is required of the findings).

If the project budget is limited, consider doing the card sorting in-house, potentially with support or mentoring from an external expert.

This should leave budget available for the two key steps: determining a draft structure (Chapter 10), and designing draft page layouts (Chapter 12).

If you are creating a brand new intranet, there will be less information to draw on when determining card sorting topics. Existing documents and manuals can be used, as well as files on shared file servers. Discussions with key stakeholders and staff can also uncover potential information for the new intranet.

In practice, it can be difficult for organisations unfamiliar with intranets to determine what should go on the site. In these cases, consider launching a small initial site to demonstrate the potential benefits of intranets, and grow the intranet rapidly after that.

In this approach, consider skipping card sorting, and proceed directly to creating an initial draft structure (Chapter 9).

Figure 9-2: A pack of card sorting cards, with topics printed onto labels and stuck onto filing cards.

Create the card sorting cards

The list of potential topics developed by reviewing current and planned content needs to be distilled down to a set of 80–120 topics for use in the card sorting session. (Any more than that and the session would go for hours.)

When refining the final list of topics:

- *Find the right level of detail.* A topic such as human resources is too broad, while rate of leave accrual is probably too specific. As a general guide, each item on the list needs to be self-contained, cover one logical group of content and have the same level of granularity.

- *Focus on uncertain areas.* Within the limits of participants' time and energy, it's worth focusing on areas that are less clear to the intranet team. For example, there will almost certainly be a section for HR, but what should be done with policies and procedures?

- *Avoid existing categories and structures.* Including cards such as 'HR policies and procedures' will naturally lead participants back to that structure, thus short-circuiting the purpose of the card sorting.

- *Avoid words such as 'manual' or 'guide' in topic names.* These reflect the current format of materials, not the actual content, and are likely to bias the card sorting session.

Once a final list of topics has been determined, write these onto filing cards. If necessary this can be done by hand, but it is preferable to print the items onto labels and then stick those onto cards, as shown in Figure 9-2. This saves a lot of time and ensures the cards are legible.

It is also useful to number each card, to assist in transcribing the results later. Do this in a small font so it is not obvious to participants, lest it cause confusion.

Note that card sorting can also be done using Post-It® notes, and this allows them to be stuck on a wall instead of grouped on a table. It is even possible to use hexagonal Post-It notes, which can aid in grouping the items. Both filing cards and Post-It notes work equally well, and the differences come down to personal preferences.

Select participants

Card sorting sessions are conducted with groups of end users (staff). Where there are a number of distinct staff groups, hold one session for each group, ensuring that staff are similar in role and seniority within the groups. This allows the results to be compared across the different staff groups and identify similarities and differences in their ways of working.

Card sorting is typically done as a small group exercise (4–8 people), although there are many different variations (see a later section for more on this).

Regardless of the approach taken, it is imperative that the participants are actual end users of the intranet, not stakeholders or managers. It is meaningless to involve people as 'representatives' of other staff, as this will not provide useful results.

Note that managers may be involved as staff in their own right, to determine their needs as users. The same can be done for executives, or other stakeholders.

Run the sessions

Allocate several hours for each session. This provides about an hour to complete the card sorting, as well as sufficient time to introduce the session and wrap up.

Start by introducing yourself and the purpose of the session. An overview should also be given of the intranet project, assuming that staff know nothing about it.

Participants are then given simple instructions for the session itself:

> "Group the cards into piles that make sense to you. You can have as many or as few piles as you want. When you come up with the piles, I'll get you to write a label for each of the piles.

> "All the cards needs to be grouped into piles. If you don't know what one of the cards means, just ask me. If there is any information missing that you commonly use on the intranet, you can write an additional card.

> "Forget everything that you've experienced on the current intranet, and just focus on grouping the cards into piles that make sense to you."

It will not be long before the participants are getting stuck into the pile of cards (Figure 9-3).

Experience shows there is a common dynamic for the card sorting sessions:

- Participants are initially overwhelmed by the number of cards, and are unsure how to start.

- Once they start working with cards, the 'easy' cards are grouped and the energy levels are high.

- If given enough time, participants may completely rework their groupings mid-way through the session.

- Participants are now left with the 'hard' cards, and the pace slows.

The facilitator of the session does little beyond observing, and answering any queries regarding the meaning of cards. Beyond that, they should not become involved in the session, or in any way guide the way the cards are grouped.

Towards the end of the session, participants can be allowed to leave any 'too hard' cards ungrouped, as this provides additional information for the intranet team. It also ends the session on a positive note for participants.

Figure 9-3: A card sorting session underway, with participants grouping the cards into piles.

Figure 9-4: Participants write labels for each of the piles, which provides the intranet team with valuable ideas about top-level navigation.

Record and use the results

At the end of each card sorting session, note down the groups made by the participants, using the numbers written on each card (Figure 9-2) to speed up the process. (This will help you to get out of a booked meeting room when the next meeting is waiting in the corridor outside.)

Once all the sessions have been completed, compare the results from each card sort, looking for patterns. In an ideal world, the labels created by participants (Figure 9-4). would be consistent between sessions, and could be directly used as the top-level navigation items for the new intranet.

This is unlikely to be the case. It is natural to find a fair degree of variation between the card sorting results, due to the differences in group composition, dynamics during the session, and from the idiosyncrasies of the technique itself.

The intranet team is therefore looking for patterns across the sessions:

- Common groups suggest likely navigation items.

- Variations may highlight important differences between staff needs and behaviour across different groups.

- Labels for piles provide valuable input into top-level menus and navigation in general.

- Confusion over the meaning of specific cards highlights areas that may need further consideration.

The evaluation of card sorting results can be done via a visual comparison, using a simple spreadsheet, or more complex evaluation such as 'cluster analysis'. If online card sorting tools are being used, these often include built-in functionality for uncovering and evaluating patterns.

It is important to highlight that card sorting is not a *design* technique. The goal is not to produce a fully formed site structure out of the session. Participants are not expected to be designers, and should not be queried on their choices, or forced to revise 'bad' decisions.

In practice, the discussions between participants during the session are as important as the results themselves. These will uncover opinions about the intranet, commonly used areas, major frustrations. Ensure that good notes are taken, and make use of these when designing the new intranet.

About half of the value comes from the results, and half from the participant discussions. For this reason, avoid devoting excessive time and resources to card sorting sessions, and don't over-engineer the analysis of the results.

Variations on a theme

There are almost as many variations on card sorting as there are information architects. These include:

- Smaller or larger group sizes.

- Running one-on-one card sorting sessions, where individuals group the cards in isolation.

- 'Remote' card sorting where participants in other locations are mailed a pack of cards, conduct the session themselves, and then mail back the results.

- Getting participants to write their own cards at the outset of the session, and then grouping those.

- More complex analysis of card sorting results, using techniques such as 'cluster analysis'.

- Online card sorting, where individuals group topics using specialist card sorting software, not unlike playing 'solitaire' online.

All of these approaches deliver broadly the same results. The only exception is 'closed' card sorting, where participants are given the top level categories, and are asked to put the cards into these piles. This is intended as a validation exercise, but it is recommended that tree testing be used instead, as outlined in Chapter 11.

Consider online card sorting

Online card sorting is being used ever more frequently, and it has the benefit of allowing a larger number of staff to be involved. The online software often has built-in reports, including cluster analysis, which can make analysing the results much easier.

There are a number of solutions on the market, both free open-source packages, and commercial offerings. Some are used on the web, while others are downloaded packages that run locally.

One service which is growing rapidly is *OptimalSort*, which provides a polished web-based interface for online card sorting:

www.optimalworkshop.com/optimalsort.htm

Other ways of understanding staff

By definition, people who design and manage intranets are not typical staff. It is therefore vital that intranet teams 'get into the heads' of the staff they are designing for. This avoids the common trap of creating a site that works well for the designer, but not for others (Chapter 3).

While card sorting is one way of doing this, it is not the only way. The needs analysis techniques outlined in Chapter 4 also provide useful approaches, including:

- one-on-one interviews

- workplace observation

- contextual inquiry

While the techniques are generally the same, the focus is somewhat different. When used as needs analysis, the goal is to uncover current pain points and potential intranet opportunities. When used as 'user research', the same techniques look in greater depth at exactly how staff interact with sites and systems.

For example, staff can be asked not just where they find a phone number, but to show exactly how they look up the details in the system. This gives insights into how they navigate the intranet, what specific functionality they use, and where they struggle.

These techniques provide very rich insight into user behaviour, as well as giving valuable context around intranet usage. Every other source of information should also be brought together before creating a new site structure (Chapter 10). This includes usage reports, and the results of earlier research.

Out of all of these techniques, card sorting can be the easiest to conduct, and the quickest to run. It doesn't, however, provide the same detail and richness as some other user research techniques.

For this reason, some professionals have dropped card sorting from their toolkit, preferring to use other techniques to uncover insights about staff. Several of the accelerated methodologies used by teams (as discussed in Chapter 16) also skip card sorting.

Regardless of the approach taken, ensure that the team has 'walked in the shoes' of staff sufficiently to be able to create a site that works well for all staff.

Outcomes

If intranets are to escape the mistakes outlined in Chapter 3, they must be designed in a way that works for staff. This includes matching the way staff navigate, the terms they understand, and the groupings of content that make sense to them.

Card sorting provides one technique for intranet teams to build a stronger understanding of how staff think about information, and how they might look for content on the intranet.

By the end of the card sorting sessions, the team will have many ideas about potential navigation items and labels for use on the intranet. The discussions during the sessions themselves will also have thrown up useful insight into intranet usage patterns and common staff needs.

This is supplemented with the results of user research techniques, such as interviews, workplace observation and contextual inquiry. Together, the intranet team now has a rich understanding of how staff work, and how they use the sites and systems that are provided to them.

These insights form a key input into the development of a new structure for the intranet, as discussed in the coming chapter. They will also be used throughout the design process.

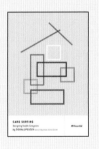

Card Sorting: Designing Usable Categories
by Donna Spencer

Need more information on card sorting? Look no further than this definitive book devoted entirely to card sorting. Every aspect of card sorting is covered, along with variations, tips and tricks.

For more information:
www.rosenfeldmedia.com/books/cardsorting/

Chapter 10

Develop a draft structure

The most common complaint from staff is "I can't find anything on the intranet". This is no surprise, as intranets can grow to be huge in size, delivering an overwhelming range of information to every corner of the organisations they serve.

This puts considerable pressure on the structure of the intranet. If it isn't working extremely well, staff will consistently struggle to find required information quickly and easily.

As a consequence of this, restructuring the intranet often becomes the primary focus of an overall intranet redesign. It is also a priority for new intranet projects.

Having completed card sorting sessions, the next step in the design process is to develop a draft structure (Figure 10-1). Note that this may be called different things, depending on who you talk to:

- intranet structure

- navigation

- information architecture (IA)

- intranet taxonomy

These all mean the same thing: where pages and documents are placed in the overall structure of the intranet, and what navigation links are provided to find them. (As an aside, 'taxonomy' actually has a much broader meaning, and it is therefore misused in this context.)

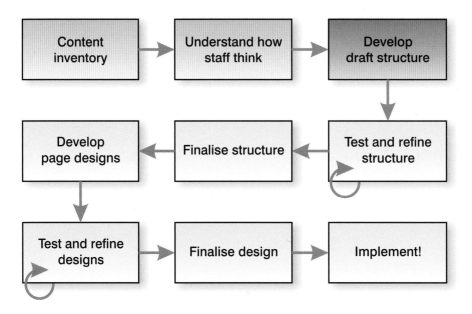

Figure 10-1: Once the needs analysis and card sorting are complete, a draft structure for the new intranet can be developed.

Make use of many inputs into the new structure

Many sources of information and insight can be used to guide the development of a new site structure. These include:

- understanding of staff needs and practices (Chapter 4)

- results of card sorting sessions (Chapter 9)

- usage statistics (most popular pages, etc)

- search engine usage reports (most popular searches, failed searches, search terms used)

- intranet strategy and business priorities (Chapter 6)

- staff satisfaction surveys and internal communications reviews

- personal and organisational experience

- common approaches taken by other organisations

- best practices and fundamental principles

When redesigning an existing site, most or all of this information should be available. Draw together every possible resource before starting into the re-structure, as this will deliver the best possible outcome with the least additional effort.

If designing a new site from scratch, there will be much less information to draw upon. This puts greater weight on the needs analysis activities, industry best practices, and experiences in other organisations.

A brand new intranet will, however, be relatively small at the outset. This reduces the pressure on the design process, and helps to reduce the amount of up-front research required.

Develop a new structure

The process for creating a draft structure for the new or redeveloped intranet is straightforward at the basic level:

1. Start with the results of the content inventory that mapped out the major sections on the current intranet (Chapter 8). If creating a brand new intranet, develop a list of planned content areas.

2. Use the results of the card sorting sessions (Chapter 9), along with the other inputs outlined earlier, to develop top-level content groups.

3. Work downwards through the intranet, creating sub-sections and below.

4. Double-check that key content items are appropriately 'bubbled up' to higher levels of the structure, and that they are placed in obvious locations.

Throughout the process, concentrate on creating sections and labels that will make sense to staff, and not just to the content owners (see below for more on this).

In due course, every page on the existing site needs to be mapped to a location on the new intranet. At this early stage of the design process, however, this is not necessary. The goal here is to determine the overall structure of the intranet, along with enough detail to allow it to be meaningfully tested with staff (Chapter 11 to come).

There are several ways that the draft structure can be captured and documented. At the simplest level, an Excel spreadsheet can be used to list the sections and sub-sections in the new structure (Figure 10-2). This is quick to produce, and is more than sufficient for testing purposes.

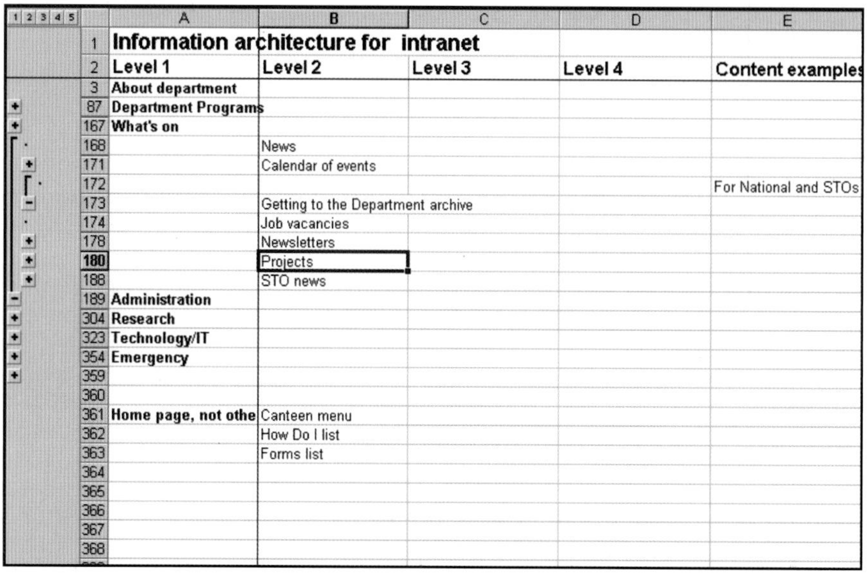

1 2 3 4 5		A	B	C	D	E
	1	**Information architecture for intranet**				
	2	**Level 1**	**Level 2**	**Level 3**	**Level 4**	**Content examples**
	3	About department				
	87	Department Programs				
	167	What's on				
	168		News			
	171		Calendar of events			
	172					For National and STOs
	173		Getting to the Department archive			
	174		Job vacancies			
	178		Newsletters			
	180		Projects			
	188		STO news			
	189	Administration				
	304	Research				
	323	Technology/IT				
	354	Emergency				
	359					
	360					
	361	Home page, not othe	Canteen menu			
	362		How Do I list			
	363		Forms list			
	364					
	365					
	366					
	367					
	368					

Figure 10-2: The draft structure for the new intranet can be quickly and easily documented in a spreadsheet.

For more complex intranets, it can be useful to create a full sitemap in diagram form, as shown in Figure 10-3. This provides a richer view of the new structure, and can show content grouping, cross-links, and content reuse.

Many different software packages can be used to create these types of site maps, including Microsoft Visio on Windows, and Omnigraffle on Mac.

A site map in this form is generally produced later in the design process, once the structure has been finalised, or as a key deliverable to hand over to developers or external agency.

Create a structure that works for staff

The fundamental goal of developing a new structure for the intranet is to produce something that works well for staff. As highlighted in Chapter 3, a common trap is to create an intranet design that makes perfect sense to the designer, but not to the staff who will be using it.

The intranet is viewed very differently by site owners (and designers) and end users (staff). When designing and managing the intranet, the whole site is laid out in its full glory. Decisions can be made on where to place a page based on

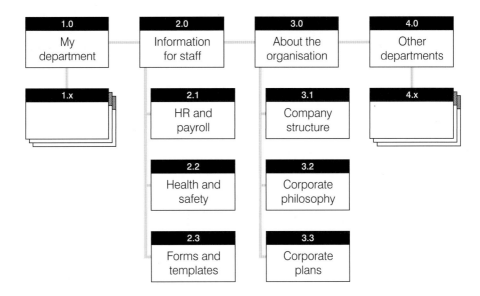

Figure 10-3: A draft site structure documented in diagram form, allowing a richer representation of the new intranet.

a global view of what's already on the intranet, as well as an understanding of the design principles than underpin the structure of the site.

End users (staff members) see none of this. When they come to the site, all they see is one page at a time, starting with the homepage. They don't know how the site was structured, or the logic that was applied when publishing new pages. Their task is to choose which link to click, based on what they read on the current page.

This disconnect can be overcome by taking a user-centred approach to structuring the intranet. This ensures all staff can use the intranet, and not just the designers and content owners.

The intranet should also be designed with an eye to the future. Over time, new capabilities and content will be progressively added to the intranet. The structure developed today should be scalable to meet tomorrow's needs, as far as they can be foreseen.

This maximises the sustainability and maintainability of the intranet. A good structure provides clear answers on where new content should go, as well as expanding gracefully as the intranet grows in use and importance. In this way, staff needs will continue to be met in the future, and not just at the instant when the site is launched or relaunched. (More on this in Chapter 19.)

Art and science

Intranet teams must draw on a mix of rules, guidelines, experience, creativity and common sense when developing a new site structure.

As discussed on page 86, much of the work involved in designing or redesigning an intranet can be done in-house by the core intranet or project team.

There are, however, several key steps in the process that can benefit from additional expert assistance. Developing the draft structure for the site is one of them.

Due to the mix of art and science involved in creating a new structure, prior intranet experience can be of considerable benefit.

When looking for a consultant or contractor, ensure they have very strong user-centred design (usability and information architecture) skills.

They should also have in-depth experience with intranets, including an understanding of intranet strategy and management.

This ensures that the end-product does not just meet industry best practices, but it will also sustainably address business needs.

There is no single 'right way' of structuring an intranet. Navigation will vary depending on the nature of the organisation, the work that staff do, the size and complexity of the site, the publishing model and a hundred other factors.

For example, Figure 10-4 shows a number of different top-level menus across a range of intranets.

While there is considerable variation, it is still valuable to look at what other organisations have done, particularly to see different solutions to common problems.

There is no recipe or step-by-step guide for creating a structure that works well for staff. There are, however, best practices and fundamental principles that can be used to guide the development of a new intranet. These light the way towards better designs, drawing upon the experiences of many intranet teams, and the input of usability experts.

This chapter will outline the most important principles that help intranet teams to improve intranet navigation, and to deliver a site that allows staff to quickly find what they need.

We will also explore common pitfalls and traps, the 'worst practice' that can bedevil intranet projects. Some approaches to intranet design simply don't work, there's no necessity for every intranet team to discover this the hard way.

The development of a new draft structure brings together all possible inputs, and guided by experience and best practices, produces something that works better. From here, user testing (as outlined in Chapter 11 to come) allows the structure to be refined and improved.

Let's now explore some of the fundamental principles that underpin a successful intranet structure.

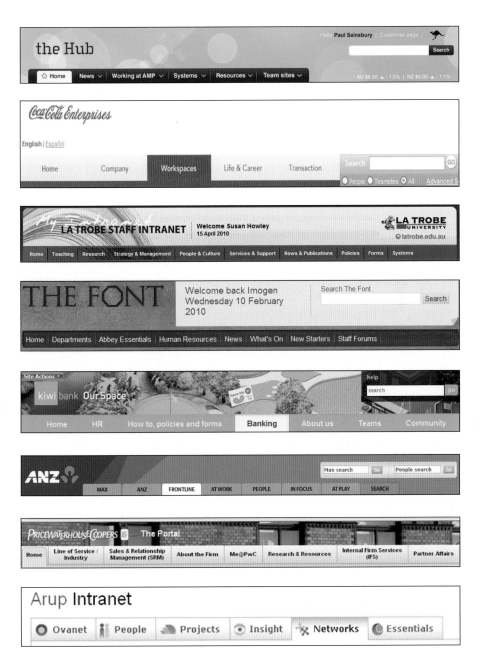

Figure 10-4: Top-level menus from a sampling of intranets, showing some similar items but many differences. From top to bottom, screenshots courtesy of AMP, Coca-Cola Enterprises, La Trobe University, Westminster Abbey, Kiwibank, ANZ, PriceWaterhouseCoopers Canada, and Arup.

Escape the org chart

As outlined in Chapter 3, one of the common traps is to structure content on the intranet according to the organisation chart. Content is owned by individual business units and teams, and it is entirely natural for them to group 'their' information in a single location. This gives them a clear place to publish to, and matches the way they work.

While this is beneficial for publishers, it is the single biggest cause of staff frustration with the intranet. It means that staff need to know who owns the information before they can find it. In large organisations this is extremely difficult. With the near-constant pace of organisational restructures, there may be few staff who understand the current org chart.

In the most extreme cases, intranets can become collections of separate intranet 'sites'; islands of content managed independently by individual business units. Intranet homepages surface key information, but little else is done to help staff find the content they need.

These types of intranets are often seen in large and fragmented organisations, demonstrating the principle that the intranet holds up a mirror to the organisation it serves. While it often isn't possible to consolidate all this into a single all-encompassing site, key content can be restructured to better meet staff needs (some of the complexities involved are discussed in Chapter 19).

The organic nature of intranet evolution naturally produces sites that are structured according to the organisational units that originally produced the content. Content is progressively added to the site as needed, and in the absence of any wholesale review and restructure, stays in the structure that was first developed.

Well-designed intranets gets away from this publisher-centric view of the content, and deliver sites that are structured around the needs and practices of end users (staff).

This means presenting information according to *subject* and *task*, rather than organisational unit. For example, a link to 'Travel information' is more useful than burying the information inside the Finance section (the most common owners of travel processes).

Note that this is a major shift for many intranets, and one that requires a substantial change in publishing processes and governance. As will be discussed in Chapter 15, this will require careful change management, extensive communication, and in-depth education.

Maximise information scent

In the book *Don't Make Me Think*, Steve Krug outlines a fundamental principle when structuring a site: make sure users know where to click to find the information they need. If users are uncertain or hesitant, the designer of the site has failed.

All staff have to go on when navigating the intranet is what's on the current page. They don't built up a mental model or map of the site, and have little recollection of where they've previously been.

Some links and labels give a clear sense of what they mean, and what they will lead to. Others give few clues, or none at all. As much as possible, intranet designers should provide staff with meaningful and helpful links, labels and categories.

Underpinning this is the concept of *information scent*. This is a term used to describe how people evaluate the options they have when they are looking for information on a site. When presented with a list of options users will choose the option that gives them the clearest indication (or strongest scent) that it will step them closer to the information they require.

The term information scent was first coined by researchers at Xerox Palo Alto Research Center (PARC). Their research revealed similarities between the way humans search for information and the way animals hunt. This was picked up by Jared Spool and his team at User Interface Engineering (*www.uie.com*) who have conducted much research in this area.

Sites with strong information scents are good at guiding users to content. Conversely, sites with weak information scents cause users to spend longer evaluating the options they have and increase the chance that they will select the wrong option, forcing the user to employ the back button.

When structuring the intranet, links and categories should be carefully chosen to be more meaningful and useful for staff. This is the single most important principle that drives the delivery of a successful intranet.

Information scent can be used as a simple test when structuring the site. Whenever a decision is uncertain, ask the following questions:

- Will this link make sense to staff?

- Does it give a clear indication of what they will get if they click on it?

- Does it help them to confidently choose between items?

Table 10-1 shows a number of examples commonly seen on intranets, dividing them into items with little information scent, and those with more. In most cases, it is the generic, all-encompassing or vague items that struggle.

Poor information scent	Better information scent
"Resources"	"HR forms"
"Staff information"	"Booking travel"
"Useful tools"	"About the organisation"
"Document collection"	"Consumer products"
"Publications"	"R&D projects"

Table 10-1: Some links and navigation items have good 'information scent', giving a clear idea of the information they will provide. Others have little or no information scent, making it hard for staff to confidently find what they need.

It can still be hard for the intranet designer to judge whether a link is meaningful for staff, and not just for the designer. The role of user testing, both tree testing (Chapter 11) and usability testing (Chapter 13), is therefore vital.

It is all too easy to fall into the trap of creating broad, all-encompassing intranet sections, as these make our lives easier as intranet designers. For example, "staff resources" allows many items to be easily placed within it.

It is important to recognise that the logic we use as designers to categorise information is not visible to staff, who only have the links on pages to go by when navigating through the site.

The structure of the site in isolation is also not enough to ensure staff can find what they need. The design and layout of pages is equally important, providing additional *context* to navigation. This will be explored further in Chapter 13.

Structure three types of content

When working on a new structure, it quickly becomes apparent that there is no one-size-fits all approach that will encompass all intranet content. Some content is relevant to all staff, while other information is specific to just a single group of staff.

A typical intranet also contains thousands of pages, and trying to fit this into a single 'ideal' structure is a daunting prospect. It is therefore useful to take a pragmatic approach to structuring the new intranet. This focuses resources on the most-used and highest-value content, and takes a progressively looser approach for other information.

When creating a new structure, consider three types of information:

1. *Core content.* Some content on the intranet is clearly relevant to all staff, addressing fundamental staff needs or covering core organisational topics. This includes typical HR tasks, IT information, key finance forms, and details on organisational strategy.

2. *'Shop window' content.* Beyond core content, there will be a large volume of more specialist information, published by areas such as product management, customer service, engineering, and support.

3. *'Back office' content.* In addition to information shared with a wider audience, business areas may also have content that is only needed within their business area, for use by their staff only.

These three types of content are handled quite differently on the intranet, and this is reflected in the overall structure of the site. In the past, clear lines were not drawn which led to many sections and pages having overlapping purposes.

For example, the landing page for the finance section might combine 'about us' information with procedures for staff, and internal reference content. This causes confusion for staff, and makes it harder for them to find what they need.

Practical approaches for each of the three types of content will be outlined in the coming sections.

Structure core content

Core content is relevant to all staff, and it typically includes information from corporate services areas such as HR, finance and IT. In practice, it will consist of a mix of policies, procedures, reference information, updates and forms.

Bring this information into the main body of the intranet, and structure it according to task and subject. Ensure this information is quickly and easily findable by all staff, and consider adding links to key content on the intranet homepage.

Figure 10-5 shows one possible way of presenting core information. Procedures, policies and tools relating to admin tasks have been brought together onto a single page. This is despite the individual items being owned by a wide range of business units.

In many ways, this core content is the highest value content on the site. It is certainly used by the widest audience, and should be managed accordingly.

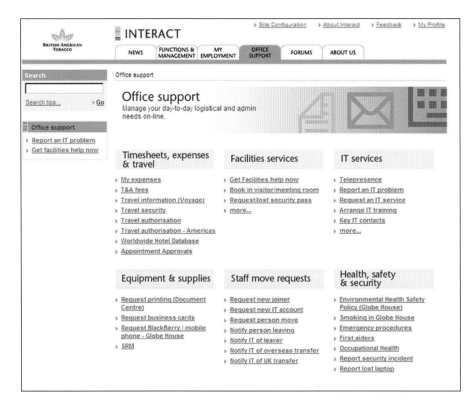

Figure 10-5: Core content should be pulled out of the organisational structure, and presented according to task and subject. Screenshot courtesy of British American Tobacco.

The central intranet team should play an active role in designing and managing this core content. While the actual information will still be owned by the relevant business unit, the intranet team needs to follow a user-centred methodology in order to deliver a single set of pages that will meet staff needs.

This can be one of the largest and most challenging changes to make to an intranet. Intranets naturally evolve according to organisational lines, based on the activities of individual publishers. Changing this is no small task, and it will often become the heart of the intranet design project.

Strong governance will need to be put in place that recognises the needs of both staff and content owners. Engage stakeholders from the outset, and use the results of card sorting (Chapter 9), tree testing (Chapter 11) and usability testing (Chapter 13) sessions to demonstrate the benefits of moving to this type of structure.

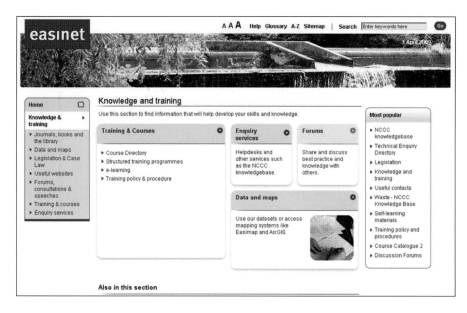

Figure 10-6: Business units often have 'shop window' content they need to share with the rest of the organisation. Screenshot courtesy of Environment Agency.

Structure 'shop window' content

Beyond core intranet content, there will still be a large amount of information needed by staff. Moving all this into a single structure is often impractical, but it nonetheless needs to meet staff needs.

A useful metaphor for content owners is the 'shop window'. This presents desired and needed items in a clear and attractive way, visible from the street outside. Careful thought should go into the design of the shop window so it is easy for current and potential customers.

Similarly, many business units publish information that is needed by staff in the rest of the organisation, whether a small number or a large number. This information must be structured in a way that can be found, understood and used by staff outside the business unit itself.

In practice, shop window content is often left in some form of organisational structure. For example, sections may be created for communications, security, facilities, and corporate planning.

Within these sections, content must be structured in a user-centred way. Figure 10-6 shows one approach to this, focusing on key information and tasks.

The section on navigation pages in Chapter 12 will explore the design of these types of pages in greater length.

Note that not every business unit needs or requires a shop window area. Mandating or pre-creating these sections in the site structure can lead to many pages of 'blah blah' content.

The central team will play a supporting role in developing the structure for these areas of the intranet. Time will need to be prioritised according to the importance of the content, and much of the work will be done by the content owners themselves.

Provide guidelines and supporting information to help business units understand the role of their sections of the site, and how to structure them for greatest effect.

Structure 'back office' content

In addition to their shop window content, many business units will also have 'back office' content. Following the metaphor from the previous section, at the back of the shop will be the door marked 'for employees only'. Behind this is hidden away all the stock and tools needed by the organisation to deliver its services.

In a similar fashion, business units will have extensive resources they need for their own use. These might include internal procedures, working documents, internal references, and confidential figures.

This needs to be clearly separated from shop window content. Even when it isn't strictly confidential or sensitive, presenting this alongside shop window content will cause confusion for staff.

There is often a large volume of this information, and separating it out can do much to visibly clean up the intranet.

Back office content may end up in a small intranet section, or in a collaboration space, as shown in Figure 10-7. It can be left to the business unit to structure the information in a way that makes sense to them. Beyond providing overall support and mentoring, the intranet team doesn't have to be actively involved in structuring or managing this content.

See the discussions in Chapter 19 on collaboration tools for more on managing this type of content.

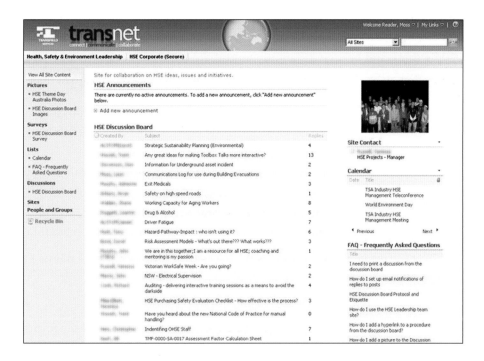

Figure 10-7: Back office content is often managed in a collaborative environment, such as this team space. Screenshot courtesy of Transfield Services.

Surface key content

Most design work starts from the top of the intranet and works down. The top level structure is determined first, and then progressively smaller sections are fleshed out.

It can also be useful to assess the intranet from the 'bottom up'. Start by identifying key items (or whole sections), based on their frequency of use or their organisational importance.

Starting with these items, work upwards to the homepage, ensuring that there is a clear path to the key items at each stage. This ensures that the important content isn't lost within larger groupings, and can help to ensure the best possible information scent.

There are some practitioners who recommend conducting the entire design process in this way, turning traditional approaches on their heads. At the very least, it is worth double-checking a draft structure to ensure that key items haven't been lost or buried.

Provide multiple ways to find content

People think and behave differently, and there will be no one structure that will make perfect sense to all staff. The way that staff navigate through the intranet will also depend on the specific task they are trying to complete.

While it is important to get the best possible structure for the new intranet, it is also important not to put all the eggs in one basket. Wherever possible, provide staff with multiple ways of finding information.

There are many ways of doing this. Cross-linking can help staff find their way to content, such as:

- *'Related links'* on pages, which cross-reference other useful content. These can be created by authors, or automatically generated based on metadata.

- *Cross-linking in navigation*, where a page may appear in multiple points in the overall site structure.

Supplemental navigation can also be included on the site:

- *Aggregating by content type*, such as bringing together a list of all the forms or policies on the intranet into a single location.

- *Index of pages*, presented according to subject or topic. (Generating an index based on page titles is of little value.)

- *Sitemap*, providing a table of contents for the site as a whole, allowing staff to quickly jump to the desired section or page.

- *'Fat footer'*, where common links are provided at the bottom of every page (at the time of writing this book, fat footers are very popular).

There are also a number of more advanced or complex approaches:

- *Metadata-driven site structure,* where the publishing tool presents navigation dynamically built out of underlying metadata or taxonomies.

- *Faceted navigation*, hybrid search and browse, and other interactive tools provided for staff to find required information.

A good rule of thumb is to avoid 'dead ends' on the intranet, so that staff always have the ability to find supporting information or further details. Start by determining the best possible 'core structure' for the intranet, and supplement this with additional links and navigation options.

In a decentralised publishing model, content owners will need to be trained on when and how to put in links. The central intranet team will also play a key role in ensuring appropriate cross-linking and navigation.

Manage 'about us' content

Business areas and teams have a legitimate need to communicate (and even promote) what they do to the rest of the organisation. Unfortunately this can become an unhealthy obsession on some intranets, overwhelming top-level pages with 'blah blah' content (see page 148 for more on this).

Some intranet teams have declared a crusade to stamp out 'about us' content, but this is unhelpful for several reasons. Firstly, the content is needed, particularly for new starters in an organisation who have no idea of who does what.

Secondly, intranet teams have little or no power to enforce their will on business units and content owners. These groups don't report to the intranet team, and rightly believe they have ownership of their own content.

While good page design (Chapter 12) can help to mitigate the negative impact of 'about us' information, it should also be considered when structuring the intranet as a whole.

In general, there are three approaches for 'about us' information:

- *About us section.* Create a top level 'about the organisation' section and place all the about information in this, structured according to the organisational chart. (The exact name of this section will vary from case to case, so make sure you test it.)

- *Include as part of the staff directory.* A good intranet-based staff directory seamlessly incorporates a dynamic organisation chart drawn out of HR systems. This can be expanded so that each business unit has a page where they can put their 'about us' information, linked to from the org chart and staff profiles.

- *Factor into shop window page designs.* If it is not possible to separate out the 'about us' information, politically or practically, carefully incorporate it into the design of shop window landing pages. For example, by making this information a sidebar, the body of these pages can still be devoted to more directly useful content.

In an ideal world, once the 'about us' content has been separated out, the rest of the intranet can be entirely task- and subject-based. This would allow the shop window areas of the intranet to move away from an organisational structure, and to be incorporated more seamlessly into the core of the site.

In practice, this transformation will be a gradual one, as significant cultural change will be required. By making sure a home for 'about us' information is structured into the intranet from the outset, a solution can be found that works for both staff and business units.

Beware of common myths

When designing intranets, it is helpful to have some rules of thumb to follow when making decisions. Over time, many of these have become elevated into principles or rules, widely used throughout the industry.

The best-known is the *three clicks rule*, which says that all content should be no more than three clicks away from the homepage of the site. This, however, is a myth.

The principle is that 'users don't like to click', and that their satisfaction with the site falls with each additional click.

This has developed into a rule that every page must be no more than three clicks away from the homepage. The big advantage of this rule is its simplicity: it's easy to state and understand, and is therefore widely known throughout the industry.

It's also perhaps the only rule that is familiar to management, outside of the intranet and design profession.

The reality is that users have no problem with clicking, as long as they are confident they're heading in the right direction. Overall satisfaction with the site is derived solely from whether they were able to find what they wanted, with little or no recollection of how many clicks were required.

This has been confirmed by research conducted by the usability expert Jared Spool (*www.uie.com*), who has carefully examined user behaviour on public-facing sites across hundreds of usability tests.

While the three clicks rule may be a myth, the common sense principle of bringing more frequently used content towards the top of the site still holds.

While users may not dislike clicking, there is no reason to make them work harder than they need to. Effort should be applied to identify common or important content, and to make sure this can be easily and quickly found on the site.

This may involve tuning the overall structure of the site, or simply adding links to the homepage.

The three clicks rule is not the only myth. The *7 ± 2 rule* (seven plus or minus two) is equally questionable. This argues that every list, including navigation links, should contain between 5 and 9 items. This includes the top level of the intranet.

As for the three clicks rule, there is some underlying common sense. Good designs are unlikely to have hundreds of links on a single page, or a top-level structure with dozens of items.

Beyond this, however, the rule starts to break down. Staff are not memorising lists, they are reading them. There are many examples where more links work better than less. For these reasons, intranet teams should avoid taking 7 ± 2 as an iron-clad rule.

There are a number of excellent articles and blog posts that explore these (and other) design myths:

www.uie.com/articles/three_click_rule/
carsonified.com/blog/design/top-10-ux-myths/
freeusabilityadvice.com/archive/35/the-3-click-rule

Search vs browse

With the rise and rise of Google, the spotlight has been shone on intranet search as a potential 'silver bullet' for meeting staff needs. Perhaps search can be provided as the primary mechanism for finding intranet information, reducing (or eliminating?) the need for intranet navigation. There is also the perennial question: do some users always search, and others always browse?

In practice, there is little evidence that some staff exclusively use search. Instead, User Interface Engineering (*www.uie.com*) findings show that on public sites at least, almost no users only search (or only browse) during usability testing sessions.

Instead, intranet designers should be guided by the following observations and principles:

- Search is often most valuable when searching for a 'known item'. That is, staff know exactly what they are looking for and what it's called, and want to go straight there.

- Navigation is most used when seeking an 'unknown item'. Staff know the intranet has something useful on a particular topic, but aren't sure what it's called or how much is available.

- Staff will often search when they aren't sure where to find something, or when they are lost.

- The worse the navigation, the more staff will rely on search.

The net result is that both search and browse are required. They must work well, and each will be extensively used by staff. Good search also relies on well-structured content, further emphasising the importance of good navigation. (See Chapter 18 for more on designing search.)

Do we still need a structure at all?

As thinking and practices evolve in the broader web space, perspectives on the design and management of intranets also change. This can lead to a number of questions:

- Do we still need to develop a top-down structure for the intranet in advance?

- Can we instead rely on more organic approaches to meet staff needs?

These questions often come from the perspective of web 2.0, enterprise 2.0 or social media, where staff are given the power to manage information directly, rather than relying on corporate approaches. This may mean the use of *tagging or social bookmarking*, or the creation of content in *wikis or other collaborative tools*.

In 2004, the term 'folksonomy' was coined by Thomas Vander Wal. This refers to end users adding 'tags' to items, which then help other users to find information. This 'tagging' potentially replaces the need for more formal metadata, and allows navigation to grow organically, driven by staff themselves.

The MITRE Corporation in the US has developed a good example of this, providing staff with a 'social bookmarking' tool for tracking and sharing useful pages. Figure 10-8 shows the homepage of Onomi, their bookmarking tool, listing recent bookmarks and common tags.

> Onomi, the social bookmarking tool developed by MITRE Corporation was commended in the 2007 Intranet Innovation Awards. For more information on what they developed, obtain a copy of the report:
>
> www.steptwo.com.au/product/iia2007
>
> The project team at MITRE have also published a technical paper on their work:
>
> www.mitre.org/work/tech_papers/tech_papers_06/06_0352/index.html

IBM has also been notable in this space, with its 'Dogear' tool providing similar capabilities. A range of vendors have also added the ability for staff to tag pages directly, and for these to be used in intranet search and navigation.

While these approaches can be valuable, they don't replace the need for a carefully developed intranet structure. Instead, they supplement and enhance existing navigation. Designers should also be aware of the experimental nature of many of these tools, and the limited adoption they have gained in some organisations.

Wikis and other collaborative tools have also proven popular in organisations, and for good reasons. They make it trivially easy for anyone to publish content, and they allow the structure and navigation to emerge organically, as a by-product of use.

This works very well in a team or project setting, where staff in the local area don't need a formal intranet structure. Team members are frequently visiting the project area, and are contributing frequently to the content and documents.

Beyond this, some have suggested that traditional navigation is dead, replaced by an organic approach where 'information structures itself'. This is overstating the case.

While wikis and other collaborative tools encourage rapid uptake, this organic growth leads to the same navigation challenges and staff frustrations experienced in other intranets. In practice, a balanced approach should be taken, which will be explored in Chapter 19.

Structuring more complex intranets

The assumption throughout this chapter is that a mid-sized intranet is being designed or redesigned. This meets the needs of a single organisation with perhaps a few thousand staff, located in a single country.

This is the most common scenario for intranet teams, and the approach of dividing content into three categories (core, shop window and back office) works well in these situations.

Some intranet teams are confronted by more complex projects:

- very large organisations (10,000 or 100,000 staff)

- complex organisational structures (such as groups of companies)

- multinational presences (multiple regions and countries)

- multilingual content

These situations add another layer of challenges onto the design process. For example, HR was included as core content that is common across the whole organisation (page 117). In a multinational firm, this is not the case, as HR policies will be defined by each country's employment legislation.

Similarly, in large organisations, shop window and back office content will be more nuanced to reflect the complexities of the org structure. There may also be additional levels of navigation that sit at the top level of the intranet.

In these situations, the simple methodology outlined in this chapter will still apply, but perhaps as only one part of a broader design process. The impact of organisational structure, business processes and culture will also have to be factored in more strongly.

These issues are discussed in greater detail in Chapter 19. There, the concept of 'global' and 'local' will help to provide a framework for designing complex intranets. If in doubt, seek help. Complex intranet projects benefit the most from professional expertise, allowing the team to draw upon experience gained across many similar projects.

Outcomes

Pulling together the results of the needs analysis with staff, card sorting results, and every other input, a draft structure for the new intranet has been developed. Either documented in a simple way (as an Excel spreadsheet) or in a fuller form (as a rich diagram), this shows the high-level grouping of content and key navigation items.

Guided by fundamental principles, and avoiding common myths, the new structure is a great improvement on the old intranet (if there was one).

The draft structure is also wrong.

The first structure developed by the intranet team is always wrong, regardless of their experience or expertise. (My work is similarly flawed.) It is virtually impossible to completely get away from a designer's perspective, or to understand every aspect of staff behaviour.

The question is: which elements of the new structure work well, and which need improvement? The draft structure could be 90 per cent right, or needing substantial reworking.

Thankfully this is just one step on the journey towards a new intranet, and in the coming chapter we'll explore *tree testing*, a hands-on technique for assessing and refining the draft structure. Using this, intranet teams can quickly improve the draft structure, and then confidently move on to the design and layout of the site.

Chapter 11

Test and refine the site structure

A new site structure has been developed for the intranet. As outlined a few pages ago, despite best efforts, this draft structure still needs work. Some aspects will work superbly, helping staff quickly find what they need. Other elements will miss the mark, leaving staff confused and lost.

The challenge is now to quickly identify problem areas, refine the structure, and repeat until the team is confident in the outcome.

There are a number of ways of doing this, but *tree testing* is one of the simplest (Figure 11-1). This provides a quick and hands-on way for the intranet team to review and improve the structure of the new intranet, before moving onto the design and layout.

This technique has been used by practitioners for years, but has gone by many different names. While card sorting is almost universally known, tree testing has been much more 'under the radar' (although this book may change that!). Credit must go to one of my former team members (Donna Spencer), for the form of the technique outlined in this book.

Tree testing is the first of the task-based tests that will be conducted on the intranet, and it involves actual staff attempting to complete actual tasks. If staff can find what they are looking for: success. If they cannot: make changes.

In only a few days, the effectiveness of the new structure can be greatly improved. While this won't be the last time the project looks at the structure, tree testing should get the team to the 80 per cent complete mark.

If time is limited, tree testing is one of the techniques that should be a priority.

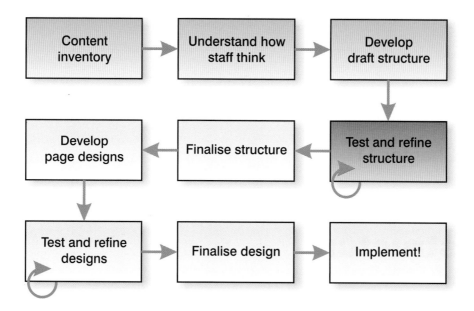

Figure 11-1: Tree testing quickly identifies weaknesses and gaps in the draft structure for the new site, and allows them to be fixed and re-tested.

Introducing tree testing

Also known as 'card-based classification evaluation', this is a quick and simple technique used to test the usability of a navigation scheme such as the structure of an intranet. It works as follows:

1. Create a mock set of menus on filing cards, matching the draft structure for the new intranet.

2. Determine a set of scenarios or tasks for staff to complete, and write these on filing cards as well.

3. Obtain a range of representative staff.

4. Get each staff person to look for the information listed on the scenarios, 'drilling down' through the mock menu structure.

5. Document where staff are succeeding (and where they aren't!), and use this to refine the navigation.

6. Write a new set of cards with the revised structure, and repeat the test.

Create the menu cards

The core of tree testing is a mock set of menus for the new intranet, written up on filing cards. These provide a simple representation of how the navigation will work on the new site, and they provide a hands-on way for staff to see if they can find what they need.

The starting point is the draft structure developed in Chapter 10. This is converted into a numbered list, as shown in Figure 11-2 below.

Starting with the top level menu items, these are numbered 1, 2, 3, etc. Underneath that, the menu items are numbered 1.1, 1.2, 1.3 … 2.1, 2.2, etc. Then 1.1.1, 1.1.2, etc.

Before long, the entire draft structure of the intranet is mapped out as a list, down two or three levels deep. (More levels could be done if required, although the value of doing so quickly becomes marginal.)

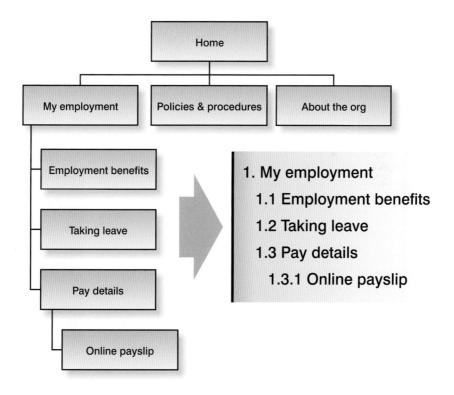

Figure 11-2: A draft sitemap for the new intranet is first converted into a numbered list, working down through the levels of the site as shown.

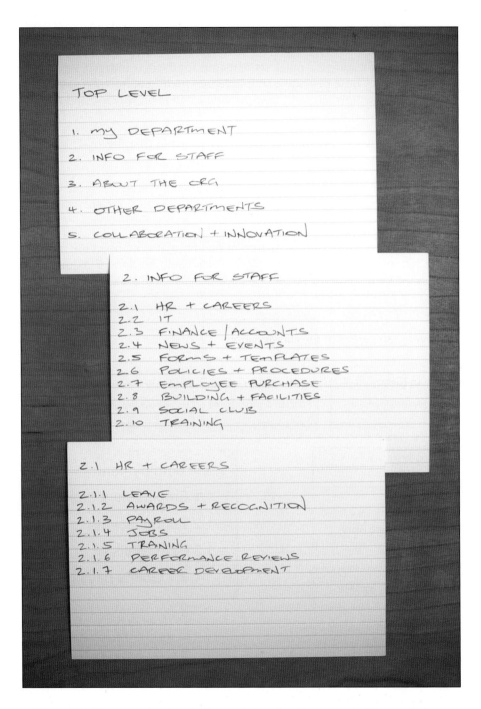

Figure 11-3: The menu structure for the new intranet written out onto filing cards, in this case, three levels down.

Once the navigation structure is converted into a numbered list, it can then be written out onto filing cards, as shown in Figure 11-3. Think of this is a paper mockup of an on-screen menu structure.

The first card shows what appears on the homepage as main menu items. Below that is the next level of navigation, written out onto multiple filing cards, and onwards down the menu structure.

Where needed, long menus can be split across several cards. The net result is a small stack of menu cards, ready for use in the testing.

Create the scenario cards

Return to the list of tasks that was identified during the initial staff research (page 43). These form the basis of scenarios for use in the tree testing sessions.

The scenarios to be used for the evaluation should:

- represent key tasks that people need to achieve using the intranet

- cover many areas of the classification

Write each scenario on a filing card in clear and readable handwriting, as shown in Figure 11-4. Alternatively, print mailing labels that can be stuck onto the filing cards. This is particularly useful if you need to create multiple sets of cards as it will save time writing.

Identify each card with a letter. This helps to identify individual scenarios when analysing the results. Keep the letters small so they do not distract participants from the actual scenario titles.

Run the testing sessions

Select a broad cross-section of staff from across the organisation, or within a single area if a targeted redesign is being conducted. Ensure that these are end users (staff) rather than stakeholders.

When inviting participants, tell them they will be performing a simple exercise to provide some feedback on the progress of the intranet design or redesign. Let them know that it will only take 10–15 minutes and that there is no preparation required.

With each participant, give them a brief overview of the project, and an explanation of the activity that they will be doing. This doesn't need to be extensive, just enough to get them working productively through the scenarios and menu cards.

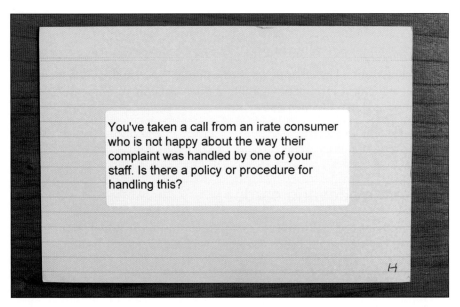

Figure 11-4: Scenarios (tasks) are written onto filing cards, for use in the tree testing sessions. Each card is labelled with a letter.

Take a step-by-step process for the tree testing with each participant:

1. Ask the participant to read the scenario.

2. Show them the top menu card, and ask them to choose where they would look for the information.

3. Show the next level down based on their selection (if they chose menu 2, show the card that has 2.1, 2.2, etc).

4. Continue until the lowest level of the menu structure is shown.

5. If the participant looks at the card and doesn't think that they are in the right place, ask them if they would like to make another choice.

6. Go back a level, or to the beginning, and follow their second choice. Don't ask them to look in more than two places as the intent is to understand where they would look first.

7. When a final menu item has been chosen, note this down against the scenario ('for scenario C, they chose 3.2.2'). If they have two attempts, note down both destinations.

The session should only take 10–15 minutes, and can quickly cover a large number of scenarios in that time.

Analyse the results

The key to tree testing is how the results are written up. Create a spreadsheet as shown in Table 11-1, listing the menu numbers down the left side (1, 1.1, 1.2, 1.2.1, 1.2.2, etc). Across the top go the tasks (A, B, C, etc).

Where a participant chooses a menu, mark an 'X'. A big x ('X') for their first choice, and a small x ('x') if they had to make a second choice.

The patterns should quickly become apparent. Task A is a problem, as there is considerable scatter. Staff weren't sure where to look, and ended up in a number of different locations.

Task B has strong grouping. In this case, staff looked in the correct location, so this is a successful result. In the case of Task C, most people looked in one location, but this was the wrong spot; a poor result.

There will rarely be a task where everyone looks in a single location, but this doesn't cause a problem in practice. As long as the majority of staff can find the majority of items, the navigation scheme is successful.

	A	B	C
1			
1.1	XXX		X
1.2	x	XXXXX	
1.2.1			x
1.2.2			XXXXx
...			
2.1		X	
2.1.1	XX		
2.1.2			

Table 11-1: The results of tree testing sessions, written up in a spreadsheet.

Refine and repeat

Based on the outcomes of the testing sessions, the intranet team can identify the areas of the navigation that need improving. This may involve minor tweaks in wording, or major restructures.

Either way, making changes to the draft structure is easy at this stage. It shouldn't take more than an hour to rewrite the menu cards, and then they are ready for use again in further tree testing sessions.

The key to this phase of the design process is that it is 'iterative'. A little bit of testing is done, and then refinements made based on the results. Some further testing, and refinement.

With a round of testing and refinement taking no more than a day, quite a few rounds of testing can be done within a week (or part thereof). Experience has shown that major fixes are made early on, so it should not be necessary to run more than two or three rounds of testing.

> There is also a variation of card sorting called 'closed card sorting', which is sometimes used to check draft intranet navigation. In this technique, instead of getting participants to create their own groups (as outlined in Chapter 9), a pre-set list of categories is provided, and participants are asked to put the cards into the category that is the best fit. The results are then assessed against the planned structure.
>
> In practice, this is much less realistic than tree testing, and gives only a fraction of the information. I would therefore always recommend using tree testing, and dropping closed card sorting from the designer's toolkit.

Benefits of tree testing

Using tree testing as part of the intranet design process provides a number of benefits:

- It provides input into the draft intranet early in the design process, when changes are quick and easy to make.

- It takes little time for the participants, who only need to give you 10–15 minutes to provide valuable input. This means that participants are simple to recruit and are happy to be involved.

- It focuses on real tasks that people would undertake. In contrast, activities such as card sorting focus only on the content, and the outcomes may not allow users to complete tasks easily.

- It is simple and easily understood. It can be explained to participants in a matter of minutes.

- It is quick and cheap to run. The evaluation takes few resources and a short amount of time so can be done as needed.

- It is low-tech and flexible. The classification can be modified during the evaluation if necessary to test alternative approaches.

- It needs no special training for organisers.

- It involves users in the design process, and demonstrates that the intranet will be created with their needs in mind.

- It allows a large number of staff to be involved in the design process.

Case study: the HR section

It can be difficult to deliver navigation and site structure that makes sense to staff in the face of the near-constant organisational restructures that occur in many businesses.

Take the case of human resources (HR). Universally recognised by staff, this business unit provides fundamental services relating to employment conditions, hiring, firing, and often pay.

In one organisation, a restructure led to a renaming of HR to 'People & Performance Services', no doubt to reflect the broader role of the new business unit. The request from this area was to rename the 'HR' link on the intranet to match.

Needs analysis conducted with staff (Chapter 4) clearly showed that staff understood 'HR', and would struggle with other names. Professional experience suggested that 'HR' would be the most effective label to use on the intranet.

During tree testing, staff consistently struggled to find information when the section was named 'People & Performance Services'. This was reinforced during later usability testing (Chapter 13). When the section was named 'HR', staff had little difficulty.

With the weight of evidence, the team returned to the content owners and discussed options. In the end, it was decided that the section should be called HR on the intranet. Most extraordinarily, the business unit itself decided to rename itself back to HR.

Fixing the navigation of the intranet through the use of these techniques is commonplace, and one of their greatest benefits is helping to cut through the swirl of often competing stakeholder opinions about navigation and naming. (Influencing the structure of the organisation itself is extremely rare, and may only be a 1 in a million outcome!)

Quantify improvements

In addition to refining the structure of the site, tree testing can also be used to quantify the improvements. Testing can be done on the existing site as a baseline which can be compared against the outcomes of each stage of testing.

There are a number of different ways of putting figures to the improvements. The simplest approach is to calculate the percentage of scenarios that are successfully completed. This might look like:

> Staff could successfully complete only 52 per cent of tasks on the old intranet. After one round of improvements this immediately rose to 75 per cent, and the final structure has a 87 per cent success rate.

If the same tasks are used for the usability testing (Chapter 13), this can allow the results to be correlated to some degree. Tree testing can also be done on the final structure, just before or after go-live.

The results of the tree testing sessions can also be shown visually, using a 'traffic light' system. A task is marked green if 80 per cent of participants can complete it successfully. A score of 60 per cent or above gets orange, below 60 per cent is marked red.

As shown in Table 11-2, the testing of an existing site could produce a lot of reds with some oranges. This quickly shifts to a mix of oranges and greens, through to the final structure which is mostly green with some oranges.

This can be a powerful way of presenting results to stakeholders.

Task	Old intranet	Round 1	Round 2	Round 3
A	52%	65%	88%	90%
B	34%	82%	82%	82%
C	65%	75%	90%	90%
D	60%	85%	85%	85%
E	48%	55%	65%	70%
F	40%	68%	52%	65%
G	78%	68%	75%	86%

Table 11-2: The results of tree testing can be shown visually, using a 'traffic light' system which summarises the levels of successful task completion.

Consider online testing

In addition to doing tree testing by hand, there is a recently developed online tool called TreeJack. This makes the testing sessions much simpler for both participant and facilitator, and generates a range of comprehensive statistics.

For more information:
www.optimalworkshop.com/treejack.htm

Simple electronic prototypes can also be created to test the structure, using a variety of tools. Some teams have even mocked up the structure on a file server, using Windows Explorer as a super-simple way of allowing staff to navigate the draft structure.

Regardless of whether the testing is done by hand or electronically, the same insights will be gained. Use an approach that you are most comfortable with, and don't be afraid to experiment over time.

Recognise the limits of tree testing

Tree testing is not a silver bullet that produces a perfect site structure for the new intranet. While the testing is reasonably realistic (staff start at the homepage and work their way down the structure), it is still testing the navigation in isolation.

On a real intranet, navigation is intertwined with page layout, with links grouped together and presented in ways that dramatically change how staff make use of the site.

Tree testing overlooks the role of cross-linking, or more complex navigation schemes such as filtered searching and faceted navigation.

Tree testing should therefore only be seen as producing an '80 per cent solution' for the navigation, with further refinement using techniques such as usability testing (Chapter 13) when the page layouts have been developed (Chapter 12).

Some practitioners skip testing the draft structure, and move straight to creating draft designs (Chapter 12). Usability testing (Chapter 13) can then be used to simultaneously test both the structure and layout of the draft site.

This can help to speed the project, and will produce a concrete design earlier in the process. The downside is that the overall structure doesn't get tested until later in the project, and the findings of the usability testing may not clearly differentiate between structure and design issues. In practice, take an approach that fits the time and resources available to the project.

Outcomes

The starting point was a draft structure for the new intranet, replete with both strengths and weaknesses. Tree testing provides a quick, simple and cost-effective way of uncovering problems with the navigation, allowing them to be tweaked (or substantially reworked).

Within a handful of days, the intranet structure has been dramatically improved to the point that the majority of staff can successfully complete the majority of tasks.

This is still only perhaps 80 per cent right, but it allows the team to move confidently on to the next stage of the design process: getting the page layouts and designs right. This will be covered in the coming chapter.

Chapter 12

Develop page designs

Having created an effective and easy to use structure for the new intranet, the team can turn their attention to the design of the pages themselves (Figure 12-1). This is a broad activity, encompassing:

- layout and functionality of pages

- design of key pages

- visual appearance and branding

Good page design is critical for the success of the intranet. Well-structured pages surface key information and help staff find what they want. They allow the intranet to expand gracefully over time, and help to address the sometimes competing demands of stakeholders.

The overall design of the intranet also sends a clear message to staff, presenting a professional image that builds trust and confidence.

Page designs must therefore encompass both core usability considerations, and broader visual design aspects. This is the 'art and science' of intranet design.

Some elements of intranet design are well understood and widely implemented. These can be copied and tweaked to fit individual circumstances. In other cases, intranets benefit from ongoing innovation, exploring new ways of delivering sites.

In part, intranet design is driven by the ongoing evolution of public websites. Innovations are often spawned on high-profile sites, and then adapted to match an enterprise context. In practice, intranet teams are encouraged to find a useful balance between best practices and experimentation.

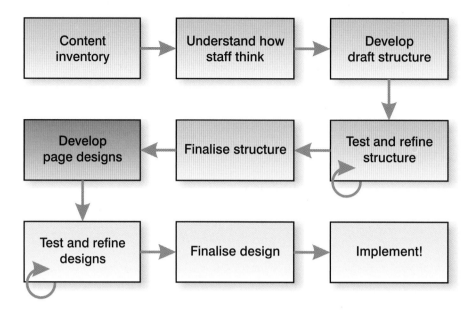

Figure 12-1: Once the structure and navigation of the intranet has been finalised, attention can turn to design of intranet pages.

Help staff complete tasks

As discussed at the outset of this book (page 18), staff visit the intranet to find a specific piece of information, or to complete a task. First and foremost, the intranet is a tool for getting things done.

This must be reflected in every aspect of intranet design. The navigation and structure must be intuitive, to allow staff to quickly get to the page they need. Pages should also be designed to give staff the greatest help to get required information, and to navigate through the site.

We will explore a number of ways that the intranet design can be shaped to help staff complete tasks:

- effective page layouts

- core usability principles

- specific design elements

- avoiding common traps and pitfalls

This vision of a 'task-based intranet' can be challenging for some stakeholders. Intranets easily fall into the trap of becoming 'publisher-centric', where information is produced in a way that makes sense to the content owners. This also generates considerable 'about us' and internal marketing content.

Creating a task-based intranet means escaping this model, moving instead to a 'consumer-centric' way of delivering content. When this is achieved, however, it is a win-win outcome. Staff can get what they need, and business units are able to publish content confident that it will be used.

Navigation versus content pages

When designing the intranet, it is useful to distinguish between two types of pages:

- *Content pages* contain the information that is being sought by staff. This includes policies, procedures, news items, FAQs, support information and manuals.

- *Navigation pages* help staff get to the content or tools they are looking for. The homepage is clearly a navigation page, along with the main business unit 'landing pages' (such as HR, Finance and IT).

These two types of pages are designed very differently, and this distinction can be very helpful for content owners and stakeholders to understand.

Design content pages

Intranets are large and content-rich. Typically consisting of thousands or tens of thousands of pages, there are a lot of *content pages*. These may be short, providing a few key details, or long and content-heavy.

To a large extent, the web industry has settled on a standard page layout for content pages for both public-facing websites and intranets. An example of this is shown in Figure 12-2.

The content sits in the middle of the page, presented in a readable font on a white background. Navigation sits outside the content, with global navigation across the top (limited to just a handful of tools in this case), local navigation on the left, and related links on the right.

At the bottom of the page, a footer provides a last updated date, details on page ownership, and some general policy links (such as appropriate intranet usage).

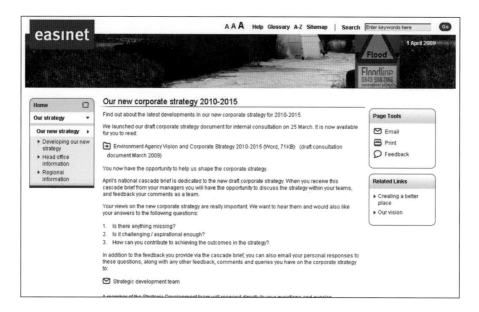

Figure 12-2: A standard layout for an intranet content page. Screenshot courtesy of Environment Agency (UK).

There are many variations on this theme, such as the design shown in Figure 12-3. This looks very different in appearance, but has similar elements to the previous example.

As long as the content pages are laid out in a plain and readable fashion, there is no reason to spend excessive time designing them. The out-of-the-box layouts provided by the publishing tool will often suffice.

Follow these commonsense guidelines:

- Ensure text is legible and easy to read, with enough contrast against the background (plain white is always a safe choice).

- Display the main heading (page title) clearly at the top of the page, and ensure headings within the body of the text are distinguishable.

- Include useful cross-links or related links, ensuring that these are directly relevant to the current page.

- Use images sparingly, and wherever possible restrict them to situations where they convey information (rather than just being decorative).

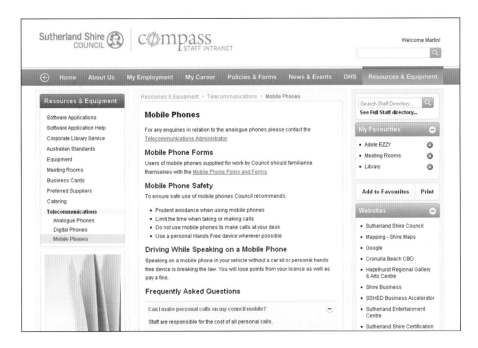

Figure 12-3: Another typical intranet content page, with global and location navigation. Screenshot courtesy of Sutherland Shire Council.

- Keep any functionality as simple as possible, or eliminate it entirely. ('Expandable-collapsable' layouts can be troublesome, for example, from a usability perspective.)

- Ensure the content itself is well-written, useful and relevant. Wherever possible it should be 'written for the web', rather than just being a cut-and-paste from a pre-existing Word document.

Design key pages

The intranet homepage is the most visible page on the site, and is by far the most used. The design of the homepage is also highly contentious, buffeted by many different stakeholder priorities and considerations. For this reason, the entire of Chapter 14 is devoted to the homepage. This leaves the rest of this chapter to explore the remaining pages on the site.

We will also explore the design of key pieces of functionality such as the staff directory and search in Chapter 18.

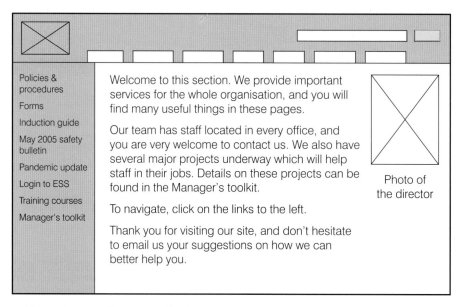

Policies & procedures
Forms
Induction guide
May 2005 safety bulletin
Pandemic update
Login to ESS
Training courses
Manager's toolkit

Welcome to this section. We provide important services for the whole organisation, and you will find many useful things in these pages.

Our team has staff located in every office, and you are very welcome to contact us. We also have several major projects underway which will help staff in their jobs. Details on these projects can be found in the Manager's toolkit.

To navigate, click on the links to the left.

Thank you for visiting our site, and don't hesitate to email us your suggestions on how we can better help you.

Photo of the director

Figure 12-4: An all-too-typical landing page on intranets, dominated by 'blah-blah' content and providing staff with little help in finding what they need.

Design navigation pages

Starting with the homepage of the intranet, the user's goal is to find the required information. If search is not used, this involves traversing the site, progressing through pages until the destination is reached.

These intermediate pages are *navigation pages,* and they are critical to the success of the intranet. To often, the homepage is intensively designed, but one click in the staff member finds themselves at a page such as Figure 12-4.

These top-level 'landing pages' within key sections such as HR and Finance are often troublesome. The navigation links are squeezed into the left-hand menu, which starts out ordered, and then ends up containing an assortment of items large and small due to ongoing organic growth.

This leaves the body of the page. As this isn't a content page, there isn't any directly useful information that can be presented. The end result is typically a page full of 'blah blah' or 'about us' content.

These types of pages work very poorly for staff. Their goal is to find a key piece of information, and this remains buried many clicks down in the navigation. These pages provide little guidance or assistance.

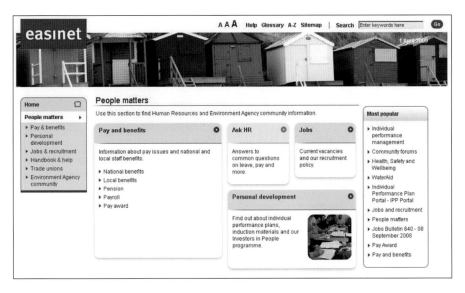

Figure 12-5: The landing page for HR (People matters) uses the body of the page to provide key links and information. Screenshot courtesy of Environment Agency (UK).

The solution is simple: use the body of these navigation pages to help staff obtain what they need. In addition to the links squeezed into the left-hand menu, the whole of the page can be used in productive ways.

There are many productive designs for navigation pages. Figure 12-5 shows one approach, where key links have been surfaced and grouped into boxes. These boxes are supplemented by a 'most popular' list, allowing staff to directly access useful resources.

Using the page in this way allows a considerable number of links to be presented, without the page becoming cluttered. This supports the goal of delivering a task-based intranet, and it works well for staff.

Figure 12-6 shows another approach to the same type of page. Even more link-rich, the careful grouping of items means that staff can quickly find what they need, despite the breadth of content. This page also escapes the perils of structuring information according to business unit or content owner, instead consolidating it into a single access point for staff.

A task-based navigation can even be taken a step further, as shown in Figure 12-7. This structures the whole intranet around five broad groupings, exposing key tasks and details, and then providing links to more traditional sections in the left-hand navigation menu.

Figure 12-6: A link-rich page which gives access to wide range of employment-related information for staff. Screenshot courtesy of British American Tobacco.

When developing these pages, start by building an understanding of the common reasons why staff visit the section of the site, including the most used documents and pages. The content owners (such as HR or Finance) will typically have a good idea of this. This can be supplemented and verified via research conducted directly with staff (Chapter 4).

Group these tasks and items in a way that will make sense for staff. Card sorting (Chapter 9) can be used on a small scale to uncover information to guide decision decisions. Tree testing (Chapter 11) can also be done within just one section of the site to validate and refine ideas.

Figure 12-7: This intranet is divided into five broad categories, each surfacing key links, tools and information. Sitting underneath this is a conventional intranet structure. Screenshot courtesy of LSI.

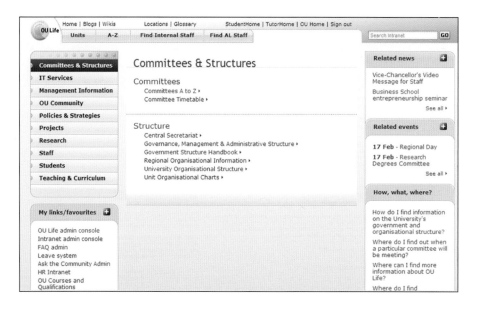

Figure 12-8: Even simple approaches to using the body of navigation pages can have a considerable impact. Screenshot courtesy of Open University (UK).

While not every page needs to be crafted with great effort, even simple designs for navigation pages can make a considerable difference. Figure 12-8 shows one approach, providing a simple list of links separated into two groups.

Taking the user-centred approach to key landing pages allow page designs to be delivered that are task-based and highly productive. It will also eliminate the majority of 'blah blah' content on the intranet.

This work doesn't end at the initial landing pages for major sections. The goal is to help staff find what they need, and attention should be devoted to navigation pages at lower levels of the site. This may involve designing pages that are three, four or even five levels down, depending on the intranet.

Prioritise design work to sections that are most used by staff. As this content is likely to be owned and managed by a business unit, the intranet team will need to play a facilitating role. Work hand-in-hand with the content owner to deliver intranet pages that work well for both publishers and users.

In addition to designing key pages on a case-by-case basis, produce a standard template and supporting guidelines for navigation pages. These can be used by content owners when creating their sections, helping to guide the intranet towards a task-based model.

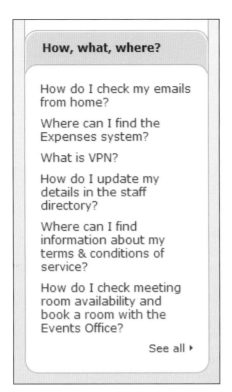
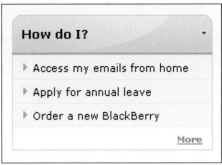

Figure 12-9: 'How do I?' provides a simple form of task-based navigation for intranets. Screenshots courtesy of Open University (UK) and Commonwealth Bank.

Use task-based design elements

In addition to designing navigation pages in their entirety, there are individual design elements that can help to deliver a task-focused intranet. The most common of these is the 'How do I?' box, as shown in Figure 12-9.

These boxes, which come in many forms, consolidate common or important tasks. When written in active language and plain English, these lists are much more effective than 'quick links' lists.

Carefully select the tasks that are highlighted on the intranet, to make sure that they accurately reflect real staff needs. (The danger is that they can express stakeholder or content owner interests instead.)

Seek out every opportunity when designing the intranet to deliver task-based elements, in small ways or as part of a larger overall design.

Is scrolling evil?

Much has been written about the dangers of long scrolling pages. From the early days of the web, considerable effort has been put into keeping content 'above the fold' (newspaper jargon referring to where the paper would be folded).

Over time further research has been put into this issue, and the results may be startling. The key conclusion is that scrolling is not inherently evil, and in many cases causes few (if any) problems.

This article provides a good summary of the current thinking:
www.boxesandarrows.com/view/blasting-the-myth-of

The summary of findings is as follows:

- If the user reaches the content page they are looking for, they are more than happy to scroll.

- If there is a clear reason and benefit to scrolling, users will do so.

- This includes scrolling on navigation pages (see the sections to come), although this may be more problematic than content pages.

- When designing pages, make sure there is a visual indication that the page is cut off at the bottom of the screen, so that staff know there is more to see by scrolling.

This does not, of course, defend long content pages filled with poorly written and rambling text. Good writing principles still hold, including the 'inverted pyramid' principle that puts key information at the beginning of the page, followed by increasingly in-depth content.

In the early days of the web, long pages utilised 'in-page navigation' with mini tables of contents at the beginning, and 'back to top' links. There are some usability concerns about this practice, and it is not encouraged where alternatives can be found.

For more information on designing pages so that staff known they should scroll:
www.uie.com/brainsparks/2006/08/02/utilizing-the-cut-off-look-to-encourage-users-to-scroll/

Staff are increasingly using a diverse range of devices to access the intranet, including mobile devices. These have dramatically different page sizes, perhaps putting an end to the concept of a single 'fold'. Staff are also being given access to the intranet from home, where computers vary wildly in terms of capabilities and resolutions.

For these reasons, pages should be designed in a way that works well across as wide a range of platforms as possible. It will not be possible to avoid scrolling on many of these devices, so it is left for us to make sure that our intranets work well in all these situations.

(The length of the homepage will be explored in Chapter 14.)

Avoid long lists

Long lists are a common trap on intranets. Examples might include a list of 100+ policies in alphabetical order, in the worst case with policies under 'P' (for 'Policy on …') or 'T' ('The policy on …').

When users definitely know which letter the desired item start with, alphabetic lists work well. (Countries of the world, states in the US, brands of cars.) In most intranet contexts, however, this is not the case. A policy on applying for annual leave could be described in many valid ways:

- Annual leave policy

- Application for annual leave

- Leave, annual

For this reason, long lists should be avoided wherever possible, in favour of breaking them up into smaller groups. The good news is that well-designed navigation pages offer solutions to the long list problem as well. The design shown in Figure 12-6, for example, works well for presenting many items without resorting to a single long list.

With the adoption of new intranet technologies, there has been a growth in on-line 'document collections', providing simple document management capabilities. These can often generate long lists of documents, which suffer from the same usability problems.

While technology solutions emphasise search and filtering options for staff, it is not clear that these are widely used or understood. Teams should therefore focus primarily on managing the length of lists, applying the same approaches used for navigation pages.

Avoid other common design pitfalls

While it is beyond this book's scope to cover every usability and design principle that relates to intranets, it is useful to address some issues commonly seen on legacy intranets.

These design pitfalls should be avoided when creating a new intranet:

- *Moving images and animations.* When endeavouring to highlight news items and updates, it is often tempting to using moving images and Flash animations. These are extremely distracting, as our eyes are programmed to always look towards moving objects, and staff have been observed to put their hands over animations in order to read the rest of the page. For these reasons, animated elements should not be used.

- *Banner blindness.* On the public web, advertising is growing increasingly intrusive. Banner ads have become a standard feature at the top of many web pages, and are creeping into the sidebar and page body. This is conditioning web users to ignore these areas of the page, and this has led to 'banner blindness'. This leads to elements in the top 50 pixels of the page routinely being ignored or missed by staff. Avoid putting key elements in this area on pages.

- *Same elements on every page.* There may be a desire to give universal access to key functionality or to broadly promote specific campaigns or links. This works fine when a few key links are included in the global navigation, but starts to erode the usability and value of the intranet when it intrudes standard sidebar elements into every page. Restrict these items to the homepage, and ensure that elements are relevant when used.

- *Too little or too much.* There is no 'right' number of items and links to include on a page. Pages that are elegantly designed to match the Apple website may be attractive, but are likely to be non-functional. Similarly, complex and cluttered pages reduce usability and use.

- *Complex functionality.* As intranet technologies improve, huge amounts of functionality become available for intranet teams to use. While well-understood by central teams, these capabilities may be unfamiliar to general staff. Limit the number of complex or technically oriented features incorporated into the site to ensure that the intranet can be used by all.

- *Reliance on icons.* Usability tests have shown that different users see very different things when they look at icons, and it is hard to come up with a set of images that will be universally understood. Intranets should therefore use icons (and images in general) sparingly, and should avoid relying on them as a sole form of navigation.

Establish a consistent design

The intranet should have a consistent design across the entire site. In practice, this means having a standard:

- overall page layout

- global navigation

- colours and branding

- design elements and behaviours

Consistency of design improves the usability of the intranet, as well as giving a more seamless experience for staff. The less staff have to adjust to differences when navigating, the easier they find using the intranet.

Business areas may want to have a distinct identity, brand or design for their sections. This is often the case when the intranet consists of a collection of separate 'sites', bound together by a corporate homepage.

It is important for business areas to understand that the intranet is not an effective marketing or promotional platform for their activities and services. As outlined on page 18, staff visit the intranet at the point of need. They are looking for a specific piece of information, or to complete a task.

They will seek out the right section of the intranet, whether it's HR, IT or Finance, and then navigate down to the required page. Outside this immediate need, they are not browsing around the intranet to understand who does what, or to see latest updates.

While a consistent design should be used across the whole intranet, it may make sense to incorporate some element into the design that allows business units to have a limited identity. If this is done carefully, it can satisfy stakeholder expectations without impacting on usability.

While there is a benefit in having a consistent intranet design, this should not become an overwhelming priority for intranet teams. There is rarely sufficient business benefit to conduct a major project just to rollout a new design, and teams should instead incorporate the new designs as one element of a larger redesign project.

Where does the intranet end? From a publisher's perspective, the intranet is often defined as the collection of pages and documents managed by intranet authors and site owners.

From a staff perspective, the intranet is much broader. Often describing it as "the thing you access via the 'blue e' icon on the desktop" (Internet Explorer), staff commonly see the intranet as anything internal that is web-based. This includes intranet applications such as HR and Finance tools.

Should these be included in the mandate for a single design across the whole intranet? In an ideal world, yes. In practice, changing IT applications is a difficult and slow process, so this should be tackled opportunistically by intranet teams.

As a longer-term solution, intranet design standards should be included in standard IT project management methodologies. Design templates should also be produced that can be used by developers when writing in-house applications.

Also include an assessment of how easily third-party applications can be 'skinned' as part of procurement processes. This could be considered as part of a broader evaluation of integration options.

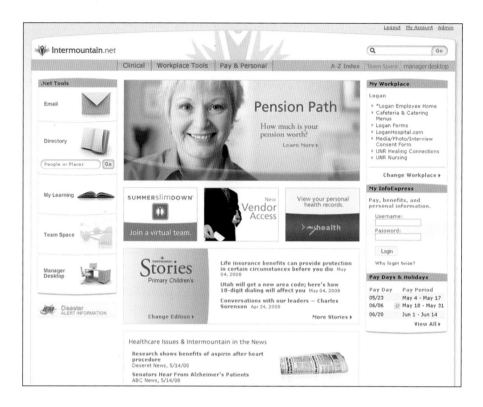

*Figure 12-10: This polished homepage design is both engaging and professional.
Screenshot courtesy of Intermountain Healthcare.*

Make the intranet attractive

The intranet cannot afford to be ugly, poorly designed or dated in appearance. This sends a strong message to staff that the site is unimportant, uncared for, and unlikely to be useful.

The visual design of the intranet should build trust and confidence in the site. It should say: 'this is a high quality site built by professionals who care about delivering an effective solution that works for staff'.

A polished and well-designed appearance shows (at least in part) that the organisation is prepared to devote resources to the intranet, and that it is an important platform for delivering information and tools.

At the most basic level, good design has an emotional impact, and this should not be discounted for intranets.

Figure 12-11: A simple out-of-the-box design has been refined to align with the corporate brand. Screenshot courtesy of Commonwealth Bank.

The primary objective must still be to deliver a useful, productive and helpful intranet. The design of the intranet does not need to be a work of art, and would typically be much more understated than a public-facing site.

But what is attractive? This is a matter of personal taste as much as professional opinion. It will also vary greatly depending on the nature of the organisation. Figures 12-10, 12-11, 12-12 and 12-13 show four very different approaches to intranet design, each with its strengths. Draw on current understanding of 'modern design' on the wider public web, and then adapt to fit the different requirements of corporate intranets.

Regardless of the design that is developed, ensure that it is visually distinct from the public-facing website. This ensures that staff know which site they're on, and whether the information is public or private. It also allows a line to be drawn around the intranet, encouraging previously separate tools (search, staff directory, HR self-service) to move within the core site.

The design should reference the corporate brand and identity, but not directly copy it. The design can then evolve over time at its own pace, reflecting corporate changes, but not necessarily in lock-step. That way, every small change to the corporate style guide won't require a major intranet redevelopment.

Figure 12-12: Intranet design that finds a good balance between the company's strong corporate branding and a clean, simple layout. Screenshot courtesy of Kiwibank.

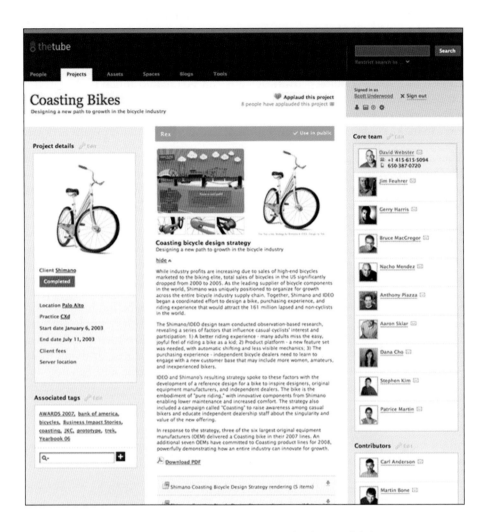

Figure 12-13: While individual elements are very simple, the overall feel is very modern, drawing on a 'web 2.0' design style. Screenshot courtesy of IDEO.

Develop draft designs

While it is possible to create designs and implement them directly into the chosen intranet technology, draft designs can be created much earlier in the process. These draft designs can be tested and refined with actual staff, as well as being tuned by the project team and participants. It is also very cheap and easy to make changes before development and implementation work is done.

These draft designs can take many forms. At the simplest end, there are rough designs drawn up by hand. These 'paper prototypes' can be vague sketches used to refine the general approach to the intranet, or more complete designs showing actual interface elements, as shown in Figure 12-14.

Known as *wireframes*, these draft designs show the layout of the page, stripped of any colours, designs or other visual elements. This focuses attention on how the page will work, and ensures stakeholders and staff don't get distracted by how it will eventually look ('I don't like that colour blue').

Draft designs can also be created electronically. These can be flat images such as Figure 12-15, which show the same detail as the hand-drawn wireframes, but with greater detail and fidelity. There are many tools which can be used to create these types of wireframes, including general-purpose drawing tools such as Visio and PowerPoint on Windows, or Omnigraffle and Keynote on the Mac (to name just a few).

In the right hands, these tools can be very productive, quickly producing a number of wireframes targeting key elements of the intranet design. This might include the homepage, major landing pages, a standard content page, and tools such as search and the staff directory.

There are also professional tools for creating wireframes, such as Axure and Balsamiq. Working prototypes can be mocked up, allowing users to click through the screens, giving a more complete experience. These tools are more complex to learn, and are mainly used by professional designers or usability experts. If you are using consultants or contractors, they may have a tool that they are comfortable with.

Finally, if the technology platform underpinning the intranet is sufficiently flexible, it may be possible to create a fully working prototype. This would typically be done later in the process, once rougher wireframes have been used to refine the key elements of the design.

This provides something very realistic to work with, but only makes sense if it can be done quickly and cheaply. If months of work and tens of thousands of dollars are required, this is excessive. If a working prototype can be created in a matter of days, and then easily changed as the design is refined, it should certainly be considered.

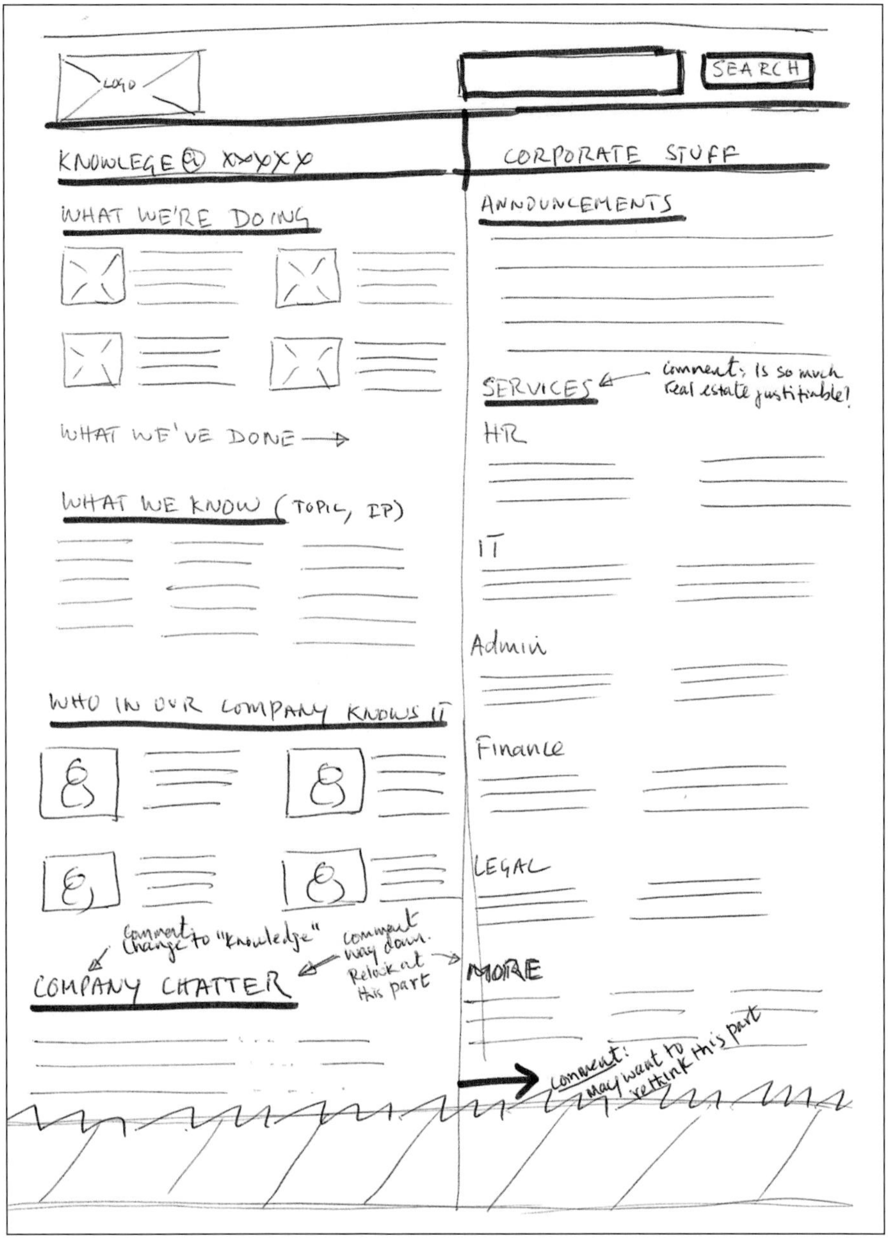

Figure 12-14: A hand-drawn sketch for a new intranet, quickly created to show key elements of the design, with annotations marking potential changes. Image courtesy of Maish Nichani.

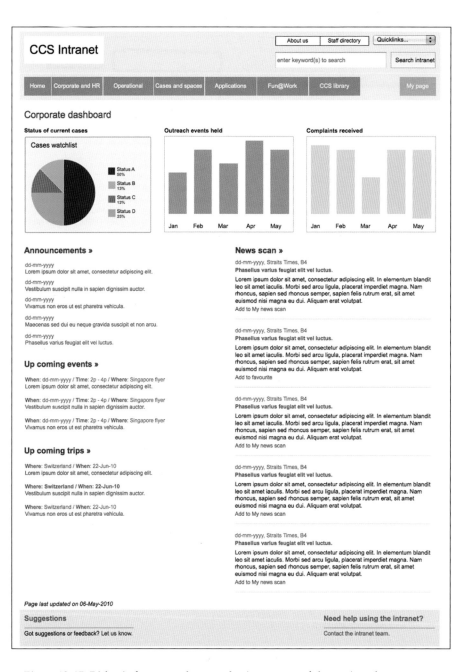

Figure 12-15: Rich wireframes can be created using a range of electronic tools, including standard drawing software. Diagram courtesy of Competition Commission of Singapore, created by Maish Nichani.

Obtaining professional assistance

As discussed on page 86, some stages of the intranet design process that benefit from external, professional assistance. This is one of them.

Great page designs draw on a broad range of usability and information architecture principles and industry best practices. They also reflect evolving thinking in the industry, as new technologies introduce additional possibilities.

While this chapter has provided a range of concrete examples, and has touched upon key principles and pitfalls, there is much more that could be said.

For this reason, there can be benefits in gaining professional assistance when developing the page designs. A good designer will draw upon years of usability knowledge and hands-on experience. They should have concrete experience with a wide range of intranet projects, and will be able to draw on past successes (and failures).

Consider taking a mentoring or collaborative approach, where the internal team works closely with the external designers. Creating a great intranet is not just a single project, and the knowledge needs to be grown in-house to allow the intranet to continue evolving in a productive way.

If there is no budget for external assistance, fear not! Designing an intranet is not a one-off project, and many improvements and refinements can be made after go-live if necessary.

Spend time looking at as many other intranets as possible, and discuss approaches with other intranet practitioners. Intranet teams are very happy to share knowledge, and this is a great support when developing new designs.

Use the draft designs

The draft designs, whether rough hand-drawn sketches or fully working prototypes, have many uses. These include:

- clarifying the overall direction for the project

- refining the scope and functionality of the new intranet

- estimating development costs

- gaining executive support and engagement

- refining the designs based on input from staff and stakeholders

- testing and improving the usability of the designs (see Chapter 13)

The earlier in the project draft designs are developed, the greater value they deliver. Having a concrete representation of what will be delivered allows everyone in the project to align their activities, and gives something tangible to work towards.

As highlighted earlier, it is easy to make changes earlier in the project, before development has commenced and resources have been committed. As the project unfolds, changes to designs should become progressively smaller, with refinements addressing smaller aspects of the new site.

Outcomes

The layout, functionality and design of the new intranet will be crucial for the success of the site. While the navigation helps staff find what they need, the design of pages helps them work out what to click on.

By drawing upon usability principles and best practices, new page designs have been developed that work well for staff and for the organisation. Starting with the top-level landing pages and working down, navigation pages have been designed to make the greatest use of screen real estate.

Content pages have been kept simple, providing readable information and relevant supporting such as related links.

By developing draft designs, in paper or electronic form, the team is now ready to progress onto usability testing, covered in the coming chapter.

Chapter 13

Test and refine designs

Having determined a structure for the intranet (Chapter 10), draft designs are created for key pages (Chapter 12). In a simple project, this may focus just on the intranet homepage and one or two other standard templates. In more complex intranets, wireframes may be produced for a wider range of pages.

Best practices were followed when creating the page designs, and careful thought was put into each element. As outlined back in Chapter 3, the fundamental challenge is that the designs will always make sense to us, as the designers. But will they make sense to our users, staff within the organisation?

Even the best designers will find it hard to avoid designing for themselves, or making assumptions about what will work for staff. Before implementing the designs, it is therefore vitally important to check with staff whether they work in practice.

Usability testing provides a very simple but highly effective way of checking designs with staff, and refining accordingly. At its most basic level, it involves meeting a number of actual staff, giving them tasks to complete on the draft intranet, and watching where they succeed (and fail!).

Note that this is very different from showing staff and stakeholders draft designs and asking them 'what do you think?'. This gathers *opinions* on the designs, but produces few insights into what will work in practice.

Usability testing produces actionable results that can be immediately used to improve designs. It identifies the 'aha!' and 'doh!' issues that could otherwise have been missed.

Best of all, usability testing can be quick and cheap, conducted by in-house team members.

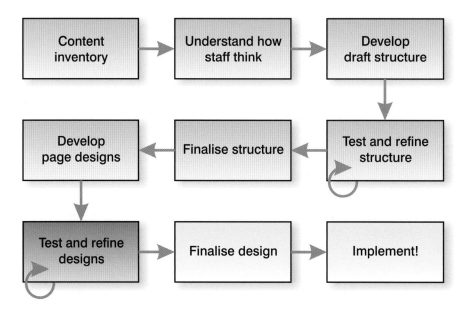

Figure 13-1: Usability testing provides a quick and effective way of uncovering problems with the draft designs for the new intranet.

Introducing usability testing

Usability testing provides a hands-on way of uncovering problems that need to be fixed with draft designs, as well as opportunities for improvement. It works as follows:

1. Gather a number of staff from across the organisation, focusing on operational staff and actual users of the intranet.

2. One by one, sit them down in front of the draft designs (in paper form, as mockups or fully working prototypes).

3. Give them tasks to complete on the intranet, ranging from simple tasks to more complex activities.

4. Observe when they fail to complete the tasks, and when they struggle.

5. Look for patterns across the tests, to uncover where improvements need to be made to the designs.

6. Revise the designs and re-test.

Rocket Surgery Made Easy
by Steve Krug

Author of *Don't Make Me Think*, Krug has recently produced a book specifically on usability testing. Focusing on a lightweight approach, this is the definitive guide for all intranet teams. A must have!

ISBN: 0321657292

Conduct task-based testing

Usability testing involves giving participants tasks to complete. These should range from simple tasks quickly completed, to challenging tasks that may be hard for most staff.

Start with the tasks identified during the initial staff research (page 43). These can be fleshed out into tasks suitable for usability testing. Simple tasks could include:

- Can you find the HR leave policy?

- Where is that news article from last week about recycling?

- Find the name of the manager of the Finance department

- Find out what staff discounts are available

More complete scenarios can also be developed:

> You have to book a trip to New York for a series of meetings. The trip will be 3 days in length. Use the intranet to book the flights and accommodation you will need.

> You are employing a new team member. Find the policy on hiring new permanent staff, and then place a job advert using the corporate system.

Aim to provide a mix of easy and hard tasks, starting with at least a few easy ones. This will give staff confidence, and will help them to relax during the test. Remember that staff are not lab rats, so we need to be nice to them during the testing!

Add extra tasks to the end of the list to cover those participants who are extremely efficient or experienced, as well as those who race through tests or abandon hard tasks early on. Having 'spare' tasks will allow time to be padded out at the end of the session if needed.

Choose what to test

As explored on page 162, draft page designs can be documented in many forms, from simple hand-drawn designs to fully working electronic prototypes. All of these are suitable for use in usability testing.

If paper prototypes are being used, recruit a team member to 'be the computer'. This involves turning the sheets of paper in response to participant's choices on where to click. With a bit of practice, this can be quite easy and seamless to manage.

If some form of electronic prototype is available, this can be usability tested. This could consist of an interactive wireframe produced by a software package such as Axure, or a fully working system quickly implemented in the chosen technology platform.

Regardless of what is being tested, ensure there are enough pages to cover the likely path of users during the tasks. This may involve developing additional pages specifically for the testing.

Set up the testing environment

Many intranet teams are likely to conduct 'budget usability testing', due to time and resource constraints. In its simplest form, this involves sitting a participant down in front of the draft designs, having a facilitator to give them tasks, and a second person with pen-and-paper to record notes.

The next step up involves recording the session. This could be done with one or two video cameras, one watching the screen and the other on the participant. There is also software that uses the built-in webcam on the laptop or PC to record the participant while recording their actions.

This software ranges from lightweight packages such as Camtasia or Silverback, through to professional-grade usability testing packages such as Morae. Prices range from a few hundred dollars to a few thousand.

Recording the session helps to ensure that vital observations aren't missed. The recordings can also be a very powerful way of demonstrating the need for improvements, and they can have a great impact on stakeholders and content owners. (A sample frame from a video is shown on page 51.)

At the top end is a full usability testing lab. Not dissimilar to a police interview room seen on TV (although hopefully more friendly!), these have one-way glass to allow observers to see the results of tests. A full recording suite is also in place.

Most larger cities will have a usability testing lab that can be hired for a reasonable daily rate. Many government agencies will have access to a public-sector lab somewhere nearby.

While more elaborate setups will provide additional information, the key findings will be uncovered even when taking the most budget-conscious approaches. If in doubt, start simple, and add more resources if required.

Run the session

Recruit staff from across the organisation to participate in the usability testing. Aim for a broad cross-section where possible, covering:

- job roles

- length of employment

- familiarity with the intranet

- IT skills

While it is generally good practice to get a reasonably representative cross-section of staff, this doesn't need to be a big deal. Experience has shown that major issues with the new design will quickly become apparent regardless of who is involved with the testing.

Welcome each participant at the outset of the test, and provide them with a brief overview of the intranet project, and the purpose of the testing. Highlight that it is a test of the *intranet*, not a test of the *user*. When they struggle to find things, this indicates a failure of the system, not a failing of theirs.

Explain the first task to them, and get them to attempt it using the prototype or draft system. Ask them to 'think out loud' during the test, explaining what they are considering and why they are choosing specific items.

This is a key element of usability testing, as it allows the team to build much greater insight into how well the intranet works, and where the problems lie. Some people will be more comfortable with this than others, and a few might need friendly reminders during the session to keep thinking out loud.

Make notes during the session on the issues encountered, even when a recording is made. (This avoids the need to re-watch many hours of test results.)

Plan on each test taking 45 to 60 minutes, depending on the participant. Ask for the participant's opinion about the site at the conclusion of the test, as well as their suggestions on potential improvements. (This is not a major part of the test, but is more about being polite.)

How many staff to test?

This is one of the most contentious questions in the usability industry. Debate has been raging for years about when the majority of problems will be found, and the cost-benefit analysis of testing with additional participants.

Some time back, Jakob Nielsen, one of the founders of usability practice, argued that testing with five participants was enough to find the majority of issues. More tests would find further issues, but on a quickly declining curve.

In contrast, findings from Jared Spool suggested that a much larger group of participants is required to find all the issues with a system. Other research has found that usability tests conducted by different practitioners on the same system find very different issues.

For most intranet and project teams, the discussion is somewhat academic. In real-life projects, the goal is to find the biggest usability issues that will cause the most problems. There is no requirement to find all the issues, as there wouldn't be enough time to address them in any case.

A smaller number of participants is likely to find the key problems, even when they aren't a completely representative cross-section of the organisation as a whole.

Test as early as possible during the project, to give the best opportunity to make changes. Instead of conducting one large and in-depth test, conduct multiple smaller tests. This allows revisions to be quickly made and then re-tested, and for usability testing to be conducted as the intranet is fleshed out and implemented.

Further reading:

www.stcsig.org/usability/newsletter/0301-number.html
www.useit.com/alertbox/20000319.html
www.uie.com/articles/eight_is_not_enough/
www.usability.gov/articles/newsletter/pubs/092006news.html

Analyse the findings and make changes

The goal of the usability testing is to find issues and problems with the new designs early enough so that changes can still be made. This is the reason why testing is done earlier in the process, on even hand-drawn rough designs, well before development and implementation has started.

When analysing the usability test, look for patterns across the tasks and participants. Where a single participant had a problem, it may sense to discount it. Where *every* participant struggled, changes need to be made.

Usability testing is very holistic, and it identifies issues relating to every element of the new site. This includes:

- placement and layout of items on pages

- labelling of links and page elements

- naming of navigation items

- behaviour of key functionality

- elements that are overlooked entirely by participants

- pages where participants are uncertain where to go next

- content that is misleading or confusing

In general, you are looking for the 'aha!' and 'doh!' moments during the tests. These will be the findings so obvious it's hard to imagine how they were missed during the initial design work.

Examples might include a button highlighted in bright red because of its importance, that is completely ignored by staff.

It might include the 'admin resources' menu containing HR and finance procedures that made perfect sense early on. During tests, however, many staff thought it was only for admin staff, and they didn't click on it.

Surprising results should be expected from usability testing, and this is where a lot of the value is derived.

As the system being tested gets progressively more realistic, the breadth of findings from usability testing increases. Tests on rough hand-drawn wireframes give useful early indications, while further tests on a working prototype provide even more information. (Both tests are useful, as early results allow changes to be made before further effort is committed to refining the designs.)

Usability testing is an *iterative* process. When issues have been identified out of the first round of testing, make changes to the prototype or wireframes. Retest to find whether these have resolved the issue, and to uncover new findings.

If a lightweight approach is taken to usability testing, it should be possible to do several rounds of usability testing and improvements within just one week. This will make a dramatic difference to the success of the new intranet, at little or no cost.

Conduct tests in-house

The description of usability testing in this chapter has been quite brief because usability testing is very simple in its basic form. Conducting simple usability tests is well within the reach of every intranet team, regardless of size or skills. Conducting tests in-house means that the design can be refined and improved even within very tight budget constraints.

This is not to say that usability professional cannot be of value. Having greater experience is extremely important when conducting more in-depth or complex usability tests. It also supports a greater depth of recommendations, drawing upon experience from past tests, and not just the current results.

External consultants or other usability professionals can play a mentoring role, helping to transfer skills to the in-house team. This can often be done alongside the usability testing, giving the best of both worlds: rapid but in-depth results, and greater in-house skills to conduct future tests.

Usability professionals can also act as a 'sounding board' for the team, helping to interpret the results of the tests, and refining recommendations and potential solutions.

At the end of the day, however, it is immeasurably better to do some testing than none! Teams should not be afraid to 'give it a go', particularly if an iterative approach is taken. By re-testing any changes made, different options can be explored without fear of making the design worse.

Outcomes

The result of the rounds of usability testing is a design for the intranet that will work well for staff. The overall structure of the intranet was refined earlier using tree testing (Chapter 11), and usability testing has now done the same for the page layouts and site functionality.

The team can now be confident that the new intranet is at least 80 per cent right. Further issues will arise as the project unfolds, and changes will undoubtedly be made during implementation. Nonetheless, by testing at this stage, many problems that frustrated and bewildered staff on the previous site have been identified and resolved.

In this way, usability testing has perhaps the greatest impact of any of the techniques outlined in the book on the success of the project and the intranet it delivers.

Chapter 14

Design the intranet homepage

The design of the intranet culminates in the homepage. There is no more contested or challenging page on the intranet. As it is the most visible page on the site, everyone wants their piece of the homepage.

There is also contention about the role of the intranet homepage:

- Should the homepage be mostly about news?

- Is navigation the key element?

- What tools should be surfaced on the homepage?

- Which links should be included?

The many stakeholders involved in the intranet will each have their own, potentially differing, opinions about these questions.

What is agreed is the importance of the homepage. It is the starting point for staff, and the jumping off point for the rest of the site. If the homepage doesn't work well, then staff frustration will quickly be heard.

There is no one 'right' design for the homepage. Depending on the overall brand and purpose of the intranet, the homepage will contain a differing mix of functionality.

Time should therefore be spent carefully designing the homepage, and getting the balance of content and tools right. Work with stakeholders and deliver a page that works well and is sustainable over time.

Seven roles for the homepage

There is a lot of information squeezed into the homepages of most intranets. This includes news, links and tools. Many of these have been added over time, as part of the overall organic growth of each intranet. Items are added at the request of stakeholders, and new applications generate further links.

The makeup of the intranet homepage is also strongly determined by the owner of the site, and the key stakeholders involved. Internal communications will naturally focus on news, while IT may concentrate on personalisation and portlets.

The intranet homepage is highly political, and intranet teams may feel powerless in the face of stakeholder demands and entrenched positions.

Many intranet homepages have also settled on a 'typical' design, even though there is little evidence that this is working well for staff.

This is an opportunity to take a fresh look at the homepage, and to use a structured approach to explore design possibilities. Use this chance to create something that is effective, intuitive and sustainable.

Regardless of who owns the homepage, there will always be contentious decisions. Key business areas may wish to promote their every activity, however minor. Some functionality may be squeezed to the edges on a homepage dominated by just a single function.

While certain conventions have built up around the design of homepages, it is not clear that these meet the needs of staff. With widespread staff complaints about difficulty in using the intranet, the 'typical' homepage design may be a contributing factor.

It is worth stepping back to take a fresh look at what goes on the homepage. This includes revisiting the way space is allocated, and the mix of functionality that is provided for staff.

In broad terms, there are seven key roles for the homepage:

1. news

2. navigation

3. key tools

4. key information

5. community and culture

6. internal marketing

7. collaboration

Each of these roles are explored in the coming pages, along with screenshots of typical designs. At the end of the chapter, a framework will be provided for deciding on the mix of functionality and how it is presented.

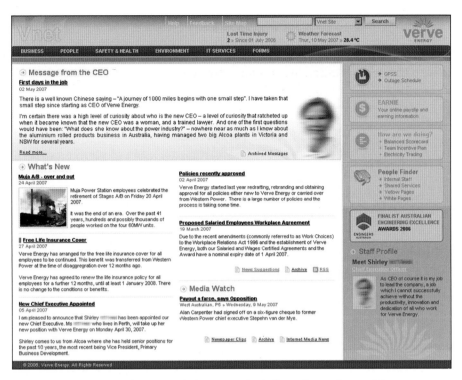

Figure 14-1: Display of news on the homepage, including a featured story. (Key areas are highlighted in white.) Screenshot courtesy of Verve Energy.

1. News

The intranet has been long recognised as a key communications channel for corporate messages. This is reflected most commonly as news on the intranet homepage, often taking up a significant portion of the overall space. (This is also the by-product of intranets that are owned and managed by the internal communications team.)

The highlighted area in Figure 14-1 shows a common design for news on the homepage. In this case, several different types of news is displayed, including corporate news and a summary of media mentions. A featured news story is also prominently highlighted.

Figure 14-2 shows another typical presentation of news, positioned prominently at the top of the page. This has been supplemented by a list of upcoming corporate events and dates, showing items that are relevant for the whole organisation.

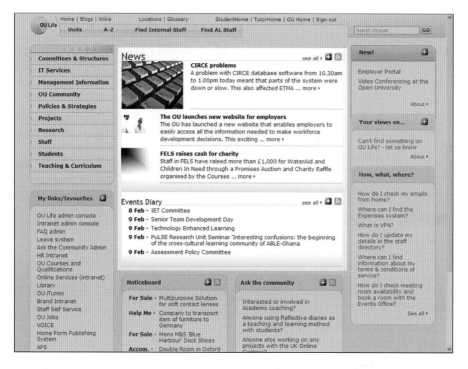

Figure 14-2: News features prominently on this homepage, along with a listing of upcoming corporate events and dates. Screenshot courtesy of Open University (UK).

As the most visited page on the intranet, the homepage is the natural place for corporate news. The intranet is also seen as a primary communications channel, one that reaches the majority of staff.

When designing the news section on the homepage:

- Minimise the number of different types of news, as this can cause confusion for staff (some intranets have crept up to half a dozen different news sections).

- Provide a well-written summary for each news item, to improve the information scent of the news section (page 115).

- Use images only when they provide additional information or context.

- Focus the news not just on corporate announcements, but also on operational updates that are highly relevant for frontline and operational staff.

- Target news to specific groups, to allow both 'global' and 'local' news to be delivered without overwhelming all staff with updates.

- Consolidate multiple news sections scattered throughout the site into a single news feature on the homepage.

- Develop a clear policy for what is published as news, to ensure that updates are relevant and useful.

Avoid 'what's new on the intranet', as staff are unlikely to pay attention to these before the point of need. Instead, write a proper news story if the update is significant enough.

While news is important, staff will not typically visit the site just to check whether additional items have been added. Communications teams should therefore focus on making the intranet useful first, with comms benefits flowing as a by-product.

2. Navigation

The primary purpose of many intranets is to act as a gateway or entry point to organisational information, and the 'first click' towards desired content. As a 'one-stop-shop' for content, good navigation is critical. This would be reflected on the homepage, exposing both the top levels of navigation and links to commonly used pages.

Despite this, there are surprisingly few intranets that devote any significant space to navigation on the homepage. In many cases, navigation is squeezed to the edge of the page, around other elements (typically news). This is, perhaps, one of the main reasons why staff are constantly complaining how hard it is to find things on the site.

In the worst cases, staff are burdened with difficult-to-use drop-down and fly-out menus, sometimes three levels deep. These generally fail to meet accessibility standards, and are cumbersome to use.

Figure 14-3 shows a robust approach to navigation on the intranet homepage. In addition to the typical 'useful links' or 'quick links' box, space has been allocated to groups of key links.

Another approach to navigation on the homepage is shown in Figure 14-4. Like most intranets, there is global navigation across the top of the page giving access to the major sections of the site. In addition, common links have been surfaced within 'our people' and 'our services' boxes.

Increasing information scent (page 115) on the intranet means providing staff with plenty of context about what links mean, and what to expect. This includes the homepage, which has a key role in helping staff choose their first click.

Figure 14-3: Navigation on the intranet homepage, providing groups of key links in addition to the typical 'useful links' box. Screenshot courtesy of Environment Agency (UK).

While it is a necessary element of intranets, global navigation is limited in how much information it can convey. Typically restricted to just a few words per link, intranet teams will often struggle to find terms that will be meaningful for all staff. (Adding drop-down menus doesn't help, and can reduce the usability of the site.)

Many intranets would therefore benefit from devoting more space to navigation within the homepage. This allows more context to be provided, and links can be grouped in a meaningful way.

Figure 14-4: In addition to global navigation across the top of the page, this homepage surfaces key links within 'our people' and 'our services' boxes. Screenshot courtesy of Concern Worldwide.

When including navigation within the homepage:

- Limit the use of 'quick links', as these can easily grow over time, and have little inherent structure or order.

- Explore the use of task-based navigation, such as a 'how do I?' section.

- Consider using space on the homepage to present key links within each major intranet section, thereby increasing information scent.

- Establish a clear policy on what links are published on the homepage, to manage stakeholder expectations (and demands!).

Intranets are used when they are useful. This means helping staff to quickly and confidently find the information and tools they need to do their jobs.

The intranet homepage provides an invaluable opportunity to surface to key information, and to directly help staff complete their common tasks. This should go beyond just global navigation or a single sidebar.

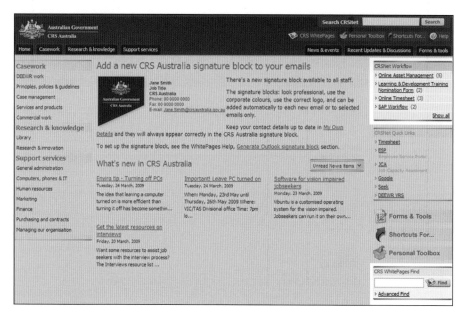

Figure 14-5: The CRSNet Workflow box provides a single list of tasks for managers, drawing from SAP and thirteen other intranet-based applications. Screenshot courtesy of CRS Australia.

3. Key tools

Tools and applications can be provided directly on the homepage, rather than just linked to. This may include core intranet functionality such as intranet search, or the staff directory. Key corporate tools may also be included if these have a relevant to a wide range of staff.

Figure 14-5 shows the homepage from CRS Australia, a mid-sized government agency in Australia. In addition to a search box within the homepage (useful for overcoming the 'banner blindness' commonly associated with links at the top of the page), a 'CRSNet Workflow' is prominently displayed.

This lists outstanding tasks for managers, and it seamlessly brings together information from SAP (via back-end integration) and 13 other intranet-based applications. This has become a key tool for staff, and it makes the intranet a true gateway for staff, beyond just a list of links to applications.

The CRSNet Workflow solution was a 2009 winner of an Intranet Innovation Award. For full details on their solution and approach, see:

www.steptwo.com.au/products/iia2009

Figure 14-6: Access to the intranet search and staff directory are prominent on this homepage, along with a view of the corporate calendar. Screenshot courtesy of EUMETSAT.

Figure 14-6 demonstrates the prominent display of intranet search and the staff directory. These are typically the most frequently used intranet tools, so it makes sense to apportion them space on the homepage.

When including key tools on the homepage:

- Include the systems and functionality that will be used widely throughout the organisation.

- Target additional tools to key user groups, such as managers, sales people and front-line staff.

- Consider automatically tailoring the homepage to present targeted tools only to the relevant staff groups.

- Seamlessly incorporate functionality without requiring additional log in ('single sign-on').

- Ensure that the functionality provided is extremely simple and usable, even for general staff.

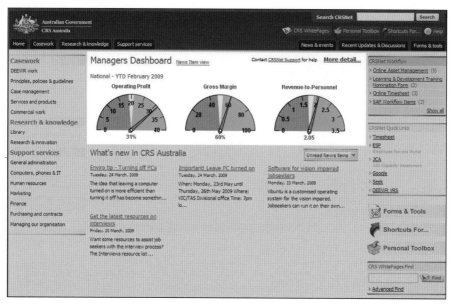

Figure 14-7: A 'managers dashboard' view of the intranet homepage provides key metrics that underpin the running of the organisation. Screenshot courtesy of CRS Australia.

The intranet can, and should, evolve into a business tool, beyond just a place for publishing content. The intranet homepage should be used to progressively deliver new tools for staff.

Be careful, however, of pursuing a technology-focused vision for the intranet. Most staff have only modest technology skills, and it is easy to deliver functionality that will not be understood or used by staff.

4. Key information

The intranet can be used to surface key corporate results and figures. Depending on the type of organisation, this might include:

- sales figures

- stock price

- safety results, such as the number of days since the last serious accident

For example, Figure 14-7 shows key metrics displayed on a homepage view for managers. These figures underpin the running of the whole organisation, and they are drawn directly out of back-end systems (in this case SAP).

Figure 14-8: Business results are displayed at the bottom of this homepage. Screenshot courtesy of Lotterywest.

Dashboard					
	Current	31-60	61-90	Over 90	All Aging
WIP	$16,825.90	$8,035.57	$3,344.80	$5,828.50	$34,034.77
WIP %	49.44%	23.61%	9.83%	17.13%	100.00%
A/R	$32,990.15	$15,445.50	$0.00	($0.02)	$48,435.63
A/R %	68.11%	31.89%	0.00%	-0.00%	100.00%
Lockup	$49,816.05	$23,481.07	$3,344.80	$5,828.48	$82,470.40
Lockup %	60.40%	28.47%	4.06%	7.07%	100.00%

Client Services

Client Search

Citrix Client Search

Figure 14-9: At this mid-sized accounting firm, the homepage displays real-time financial figures for partners and senior managers. Screenshot courtesy of Fuller Landau.

Similarly, Lotterywest displays their key business results on the homepage in Figure 14-8. At Fuller Landau, the Platinum winner of the 2009 Intranet Innovation Awards, key financial figures are displayed in real time for partners and senior managers, as shown in Figure 14-9.

Figure 14-10 shows another example of key figures on the homepage. In this case, the lack of back-end integration requires the figures to be entered manually, but the benefits far outweigh the effort involved.

Figure 14-10: Key figures can even be entered by hand, in this case each month, even where integration is not possible. Screenshot courtesy of Inchcape.

When delivering this type of top-line information, ensure that it is relevant and useful for a large cross-section of staff. The information should not get in the way of other roles for the homepage, such as news and navigation.

If the homepage is tailored for specific audience groups, it can also be used to deliver figures relevant to just those audiences.

Presenting these types of figures can give much-needed visibility to key performance indicators for the organisation as a whole. There can also be a cultural benefit in presenting these figures, such as highlighting the organisation's safety record.

Work with senior management to determine the details that best reflect corporate strategy and directions. Wherever possible, it is obviously desirable to establish back-end integration to present the figure automatically.

With the evolution of software development tools, including approaches such as 'web services', this type of integration is now significantly easier. It should not require a heavyweight portal product, and integrated figures can be displayed on even a hand-edited homepage with the right coding.

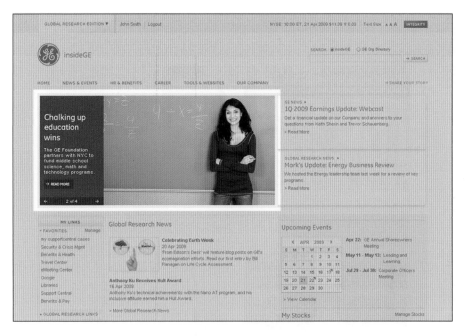

Figure 14-11: A significant portion of homepage real estate is devoted to a headline story that reinforces the GE corporate culture. Screenshot courtesy of GE.

5. Community and culture

Beyond just being functional, intranets should reflect the culture of the organisations they serve. They can also support or reinforce cultural change efforts, helping to establish new attitudes and working practices.

Organisations vary immensely in their cultures, practices and histories. This leads to many different approaches to conveying the culture on the intranet.

At GE, a significant proportion of 'above the fold' space is devoted to a headline story that reinforces the corporate culture, as shown in Figure 14-11. IDEO is one of the world's best-known design and innovation firms, and their intranet reflects this. Figure 14-12 shows how much of their homepage is devoted to showcasing innovative projects, celebrating their successes and helping to share knowledge internally.

A sense of community can also be supported more informally. The homepage of 'Boris', City of Casey's intranet (shown in Figure 14-13) is certainly distinctive. This strong branding and personality is reflected throughout the site, matching the friendlier and more informal culture of local government.

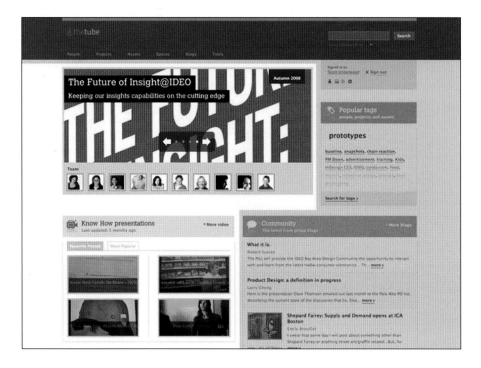

Figure 14-12: This homepage reflects IDEO's status as an innovator. Screenshot courtesy of IDEO.

Social elements can also be incorporated into the homepage design. Figure 14-14 features a typical staff noticeboard (also known as a 'buy-and-swap' area). Along with a community discussion board, these elements help to foster a culture of sharing and collaboration.

When planning cultural elements on the homepage, start from a definition of the organisation's internal brand, often determined as part of an internal communications strategy. Assess how the intranet should reflect this, and design elements of the homepage accordingly.

Where the organisation has a very strong brand and identity (such as a global consumer goods company), this should also be reflected in the design of the intranet. While not a direct copy of public-facing advertising and marketing campaigns, the intranet should have a strong design in its own right.

While the intranet alone cannot change the corporate culture, it can act as a channel for change management programs, or more subtly reflect the directions desired by senior management. Any major cultural change project should plan on using the intranet as a primary form of communication.

Figure 14-13: Boris is the star of this intranet, and is reflected in the branding and design throughout the site. Screenshot courtesy of City of Casey.

When including community and cultural elements:

- Work with senior management and key stakeholders to identify what messages need to be communicated to staff.

- Identify which elements of the corporate identity to use on the intranet.

- Establish a long-term intranet identity and branding, that will provide continuity even when public-facing marketing changes.

- Consider showcasing staff profiles, ideally with a photo (humans are hard-wired to respond to faces).

- Put a 'human face' to the intranet that goes beyond a dry corporate presentation of information.

- Find the right balance between 'softer' cultural elements on the homepage, and more functional aspects.

- Include cultural elements throughout the design of the whole site, rather than just focusing on a single element on the homepage.

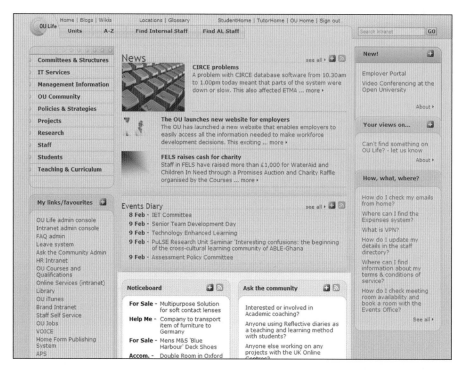

Figure 14-14: Staff noticeboards and community discussions can foster a culture of collaboration and sharing. Screenshot courtesy of Open University (UK).

6. Internal marketing

Organisations are busy places, and there is a constant stream of projects and campaigns being launched. These need to be communicated to staff, and the intranet is often a key channel.

Intranet teams are constantly bombarded with requests from business units to feature these projects or initiatives. These may be major changes, or incremental updates. They will also vary in their relevance to the broader staff audience.

Regardless of the nature of the messages to be communicated, there is a consistent requirement to promote these updates on the intranet homepage.

Care needs to be taken in how these internal marketing messages are highlighted on the homepage. Too easily, these promotional items can multiply, taking up significant space on the page. This can impact on the overall usability of the site, and the space remaining for other elements.

Figure 14-15: Internal marketing items are communicated in well-defined areas of the homepage, alongside other page elements. Screenshot courtesy of Intermountain Healthcare.

Figure 14-15 shows a good approach to internal marketing messages. A clear space is allocated to promotional items, and this is well integrated into the rest of the page design.

When delivering internal marketing on the homepage:

- Create clear areas for internal marketing messages.

- Allow some or all of these marketing messages to have an associated graphic (as in the screenshot above).

- Tightly define where internal marketing can appear, and limit the number of items that can be displayed.

- Avoid including marketing items in 'quick links' or other similar lists, as these can easily become permanent inclusions.

- Limit the time that promotional items can remain on the homepage.

- Create a clear policy for what can be promoted on the intranet.

Some organisations ask business units to 'book' space on the homepage. The items are then only up for a week, and afterwards are moved into the normal navigation.

Always look for these types of 'win-win' approaches to managing internal marketing requirements.

7. Collaboration

Supporting collaboration has become a key priority in many organisations in recent years. Fuelled by the spread of new technologies, such as blogs wikis and team spaces, collaboration has prospered.

In many organisations, it is not long before staff members are part of a dozen different collaboration spaces. This can make it very hard to keep track of the spaces, and to remain aware of updates and additions.

While the need for collaboration tools is clear, organisations are still exploring the best approaches to establishing, supporting and governing these tools. This includes determining the role of the intranet with respect to collaboration.

At worst, collaboration tools sit alongside intranets, with no connection between the two. In these cases, intranets and collaboration can end up competing, each offering a homepage and rich functionality. This lose-lose situation is to be avoided at all costs.

There are many approaches to integrating collaborative capabilities into the intranet. At the simplest end, the intranet should provide a gateway into the tools used by staff. Figure 14-16 shows one approach.

In this case, the intranet displays a list of the collaboration tools and spaces that the staff member belongs to. This displays automatically, drawn from the various source systems. This eliminates the need for an entirely new 'my spaces' navigation item or feature, and gives immediate access to currently active tools.

(Note that this list is automatically generated based on the staff member's login details, and doesn't require the individual to manually add items to the list.)

Going one step further, Figure 14-17 shows how an 'activity stream' can be incorporated into the homepage. Bringing together status updates and posts to collaboration spaces into the one area, this helps staff to see what is happening at any given point.

This can be a useful way of giving visibility to the use of social tools within the organisation, and can help to increase adoption and use.

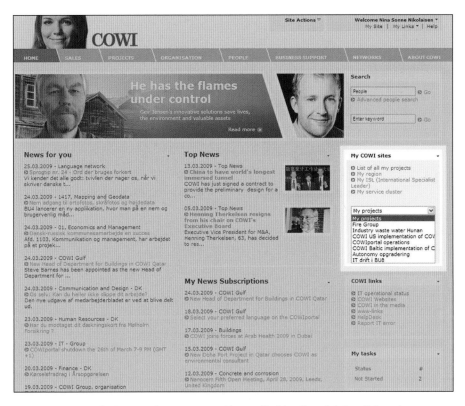

Figure 14-16: The intranet homepage provides staff with a list of their collaboration spaces. Screenshot courtesy of COWI.

Beyond this simple integration of collaboration features into the homepage, some sites are becoming 'social intranets'. These are founded primarily on social and collaborative capabilities, and reflect this through every aspect of the site.

Figure 14-18 shows one example. These intranets aim to 'democratise' the act of publishing, allowing any staff member to post news or make updates. A very different experience to 'traditional' intranets, this is reflected in the design of the homepage.

As yet, these types of intranets are relatively uncommon, and often restricted to smaller organisations. Over time, however, we may see more of these features and philosophies introduced into more conventional corporate intranets. Whatever the approach taken, endeavour to keep the design simple for staff. Complex functionality, including personalisation, will often struggle to gain uptake.

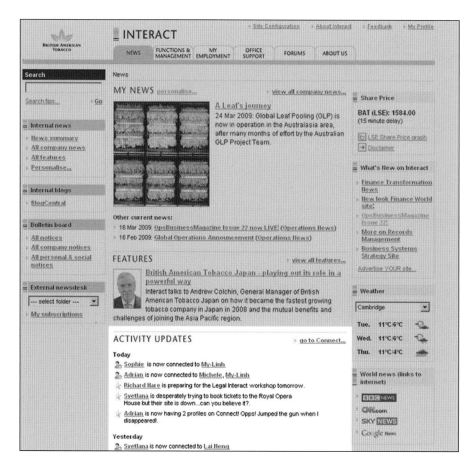

Figure 14-17: An activity stream on the homepage shows updates in the social tools and collaborative spaces. Screenshot courtesy of British American Tobacco.

Care should also be taken not to blindly replicate designs used on the broader web. While Facebook's design is well-suited for its audience, usage patterns are likely to vary greatly from those in organisational contexts. While elements can be taken from these types of sites, they should be adapted to fit staff needs. There must also be a clear purpose for their use, not just for the organisation, but also for the individual.

Intranet teams should also aim to bring intranet publishing and collaboration tools together over time. This reduces the fragmentation of information, and helps to ensure that staff don't have to look in a dozen different locations for what they want.

Figure 14-18: Homepage of a 'social intranet', emphasising broad participation and social elements, alongside traditional publishing models. Screenshot courtesy of Oxfam America.

Find the right balance

Reading through the list of functionality that can be included on the homepage, intranet teams may immediately ask: but how can we fit this all in? While many intranets to date have only included a few elements, seven different roles for the homepage have been outlined in the previous pages.

As discussed at the beginning of the chapter, homepages also typically reflect the overall ownership of the intranet, as well as organic decisions made over time.

The current design may therefore be well-entrenched for the homepage, and competing stakeholder priorities may make it hard to adapt or change the mix of functionality.

In practice, there is no one 'right' design for a homepage, and the various roles need to be balanced.

The intranet team should start by gathering together key stakeholders and content owners. Use the product reaction cards to determine the overall brand for the intranet, as discussed in Chapter 5. This will provide a foundation and framework for decision-making.

When allocating space on the homepage, make every element work as hard as it possibly can. Many features can serve two purposes, reducing the pressure on available space.

For example, a buy-and-swap area supports collaboration and culture. A calendar of events is news, as well as a tool in its own right if the events are relevant enough for staff.

Internal marketing areas can also support cultural goals if the right campaigns are promoted.

As a group, explore the potential roles for the intranet homepage. Use the intranet brand to guide the discussions, thereby helping to get a fresh perspective on the design and functionality of the page.

Table 14-1 provides an example of how the intranet brand can influence the allocation of space on the homepage.

The stronger the overall governance and ownership of the intranet, the easier it will be to gain a productive outcome. At all times, avoid the 'x versus y' arguments that are common around the homepage. There is not a single choice, for example, between communication and navigation. Both are required.

It may be useful to get participants at sessions to prioritise the elements based on their requirements (giving each person five 'points' to spend is a common approach). Bringing this together as a group help a consensus to emerge.

The design of the homepage must also be founded on strong usability and information architecture principles, as outlined in Chapters 10 and 12. This ensures that both staff and stakeholder needs are met.

Type of intranet	Sample of words chosen for intranet brand	Prioritised functionality on the homepage
Communications-focused	Engaging Fresh Relevant Responsive Trustworthy	News Internal marketing Community and culture
Content-rich	Comprehensive Consistent High quality Usable	Navigation Key tools Key information
Task-based	Efficient Essential Time-saving Useful Valuable	Key tools Navigation Key information
Collaborative or social	Collaborative Connected Flexible Personal Stimulating	Collaboration Community and culture News
Culture-focused	Appealing Attractive Friendly Fun Inspiring Motivating	Community and culture Internal marketing News

Table 14-1: Use the overall intranet brand (Chapter 5) to guide decisions about how to allocate functionality and space on the homepage.

How long should the homepage be?

The length of web pages is a hotly contested topic, and never more so than on intranet homepages. Harking back to the early days of the web, there is often a strong desire to keep the entire homepage 'above the fold' (a single screenful).

While the size of monitors and devices now varies greatly within most organisations, there is still a general desire to keep all the important content as short as possible.

When it comes to redesigning the homepage following the process on the previous pages, the difficulty of fitting everything into a single screenful quickly becomes apparent. The result is that many homepages are dominated by a single element (such as news), with a few other items squeezed into the margins (such as navigation and key tools).

We know there is widespread staff dissatisfaction with intranets, with complaints focusing specifically on how hard it is to find required information. Perhaps the restrictions of fitting the homepage into a single screen is contributing to this?

On the web, online newspaper sites have led the way in producing hugely long homepages (often in excess of 9,000 pixels!). Research by these organisations has shown that their homepages are being consistently used all the way to the bottom.

While it is never sensible to simply copy public-facing practices to the intranet, it does raise the question: how long should the intranet homepage be? Surprisingly, there is little research that strongly states that homepages should be short or long.

Figure 14-19 shows one well-designed homepage that feels very natural despite its length. Even a smaller increase in length beyond a single screenful can allow for much greater useful content, as shown in Figure 14-20.

This is left as an open question for intranet teams to explore, and they are encouraged to do so. If we can make long homepages work, it opens up the design to incorporate much more functionality, making the intranet more useful.

If this is to work, the lower elements of the homepage must be directly useful in their own right, not just an ever-scrolling list of news items or links. Good design will be needed to group items, and to 'signpost' the functionality that is provided.

Usability research has also shown that elements should be deliberately cut off at the bottom of the screen, to give a clear visual indication that there is more information available (staff can often overlook the scrollbar).

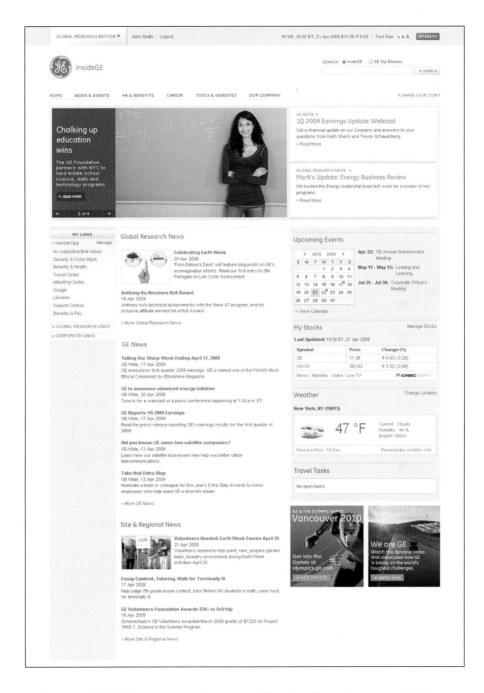

Figure 14-19: This homepage provides a range of functionality and content, and feels natural in its design despite its length. Screenshot courtesy of GE.

Figure 14-20: Even a modest increase in length beyond a single screenful can greatly improve the amount of useful information that can be provided on the homepage. Screenshot courtesy of Industry & Investment NSW (formerly NSW Department of Primary Industries).

Outcomes

The homepage is the entry point to the whole of the intranet, and the most used page. It is also the most visible manifestation of the intranet, and the most hotly contested real estate.

Intranet homepages should find a productive balance between seven different roles:

1. news

2. navigation

3. key tools

4. key information

5. community and culture

6. internal marketing

7. collaboration

With a foundation of good usability principles, intranet teams can make use of the intranet brand to determine the right mix of functionality and content. Bringing together key stakeholders and content owners, a productive consensus can be determined that meets both staff and business needs.

Chapter 15

Deliver and launch

Now that the design and structure of the intranet have been finalised, plans can be made into reality. In most cases, this will involve implementing the new intranet on some form of technology platform, such as a content management system or collaborative tool.

Depending on the nature of the technology products involved, this may be a simple matter of loading the designs into the system. Alternatively, it could be a IT-heavy project that involves months of development.

While the technology aspects are out of scope for this book, there will be plenty of other work required to get the intranet ready for go-live.

Regardless of how well-planned the intranet project is at the outset, the weeks leading up to the launch (or relaunch) of the site will always be crazily busy. While project plans make design projects look simple and linear, the reality is that many things are happening in parallel throughout the project.

This is particularly the case in the mad dash for the finish line, when a hundred small tasks and dozens of large activities need completing. Hopefully, intranet team members have been kind to their families, as they are often required to spend long hours at work in the final weeks.

To help with the planning of the project, Table 15-1 provides a list of common tasks for intranet projects. Use this as a starting point for further planning, and add items to the list as the project progresses.

While it is beyond the scope of this book to cover all aspects of intranet project planning and management, a few key topics have been highlighted in the pages to come.

Project delivery checklist

Strategy

❑ Establish project team

❑ Understand staff and business needs

❑ Develop intranet strategy (if appropriate)

❑ Determine project scope

❑ Build project support from senior management, stakeholders and others

❑ Develop business case (if required)

❑ Finalise launch date

❑ Finalise project plan and team member responsibilities

❑ Establish or refine intranet governance

❑ Determine ongoing (post-launch) resources and responsibilities

Design

❑ Determine overall site structure (information architecture)

❑ Determine metadata and supplementary navigation

❑ Finalise intranet branding and identity

❑ Develop page designs

❑ Design intranet homepage

❑ Develop page templates (in HTML)

❑ Test page templates against design standards and compliance requirements (such as accessibility)

❑ Implement designs on the chosen intranet technology

❑ Test intranet designs on all desktop environments used across organisation

❑ Test designs on mobile devices (if supported)

Table 15-1: High level outline of steps needed to deliver a new or redesigned intranet, grouped by type of activity.

Content

- ☐ Conduct content inventory and determine content ownership
- ☐ Establish authoring community
- ☐ Conduct content review and cleanup
- ☐ Determine authoring and publishing models for new site
- ☐ Provide support and training for authors
- ☐ Migrate existing content to new site
- ☐ Delete or disable old site (or old content if the two sites will be run in parallel)

Change and communication

- ☐ Develop launch strategy and communications plan
- ☐ Communicate to stakeholders throughout project
- ☐ Determine intranet name (if appropriate)
- ☐ Produce marketing materials (eg posters, displays and promotional items)
- ☐ Launch the new intranet
- ☐ Train end users (staff)

Technology

- ☐ Obtain required IT support (internal and/or external)
- ☐ Select and implement new intranet platform (if required)
- ☐ Select and implement new search engine (if required)
- ☐ Develop or refine key intranet functionality and applications
- ☐ Load test for expected day one usage
- ☐ Test intranet across organisation, covering all locations, environments and desktop configurations

Table 15-1: High level outline of steps needed to deliver a new or redesigned intranet, grouped by type of activity.

How long will the project be?

Intranet projects are often longer and more involved that most teams expect. While there is often pressure from stakeholders to deliver a new site quickly, it is important to have a realistic project plan and to set appropriate expectations.

Perhaps the simplest situation is creating a brand new intranet. In some cases, senior management gives the team just three months to deliver a new site, but this is too short. Even assuming an out-of-the-box technology is put in place, there is only enough time to put in place standard 'boilerplate' content, rather than staff and business-focused material.

In practice, even a comparatively small project to create a new intranet will likely take six months. This includes spending time at the outset of the project understanding staff and business needs (Chapter 4). A larger project that includes a greater technology component and more ambitious objectives can easily take 6–12 months.

Intranet redesigns are often more complex, due to legacy issues and the volume of content on the current site. Looking back at the four sample project scopes outlined on page 72, the timeframes could be as follows:

Incremental improvement (fix key issues with current site)	6 months
Core intranet redesign (improve site structure and design)	9–12 months
Major intranet redevelopment (redesign and new technology)	12–18 months
Intranet as a business tool (improve processes and functionality)	6–12 months

While these are just indicative figures, they may help to set broad expectations for the project. In practice, experienced teams with more resources may complete projects more quickly, while technical and political complexities will undoubtedly push the figures out.

Some intranet redesigns have stretched out to two, three or even four years, but this is clearly too long! Organisational patience will be stretched to the limit, and the drawn-out nature of the project will often mean that improvements are dated by the time they launch.

Where possible, break intranet projects into a series of activities, each producing clear business value and concrete deliverables. This mitigates project risks, and ensures steady delivery of improvements and new functionality.

Migrate content

When the project is redesigning an existing intranet, substantial amounts of content will need to be migrated into the new structure. If the project also involves moving to a new technology platform (which is common), then every page on the existing site will need to be migrated.

As discussed in Chapter 8, a primary goal of most intranet projects is to improve the quality of content on the site. There is certainly no point in launching a redesigned and restructured intranet if the content is still out of date and incorrect!

It is therefore important to carefully plan the content migration process. There are three main approaches that can be taken:

1. *Automated migration*, where software tools transfer content from the old to new sites.

2. *Migration by hand*, where content owners manually cut-and-paste content into the new intranet.

3. *Partially automated migration*, where software aids the review and transfer process.

The first option is obviously the most attractive one, but it is unlikely to be practical if any major changes have been made to the structure or design of the intranet. An automated migration also works best when both the old and new technology platforms are similar in design, and provide built-in support for importing and exporting content.

In practice, most of the content will be migrated by hand, with a few more structured areas migrated automatically (at least to some degree).

Start with the content inventory conducted at the outset of the project (Chapter 8), and develop a map for each piece of content. Ensure that only good content is migrated, and only migrate pages that have an owner.

Resist any pressure to simply migrate content quickly and to clean it up afterwards. Human nature means the clean up will never occur, and this is the single best opportunity to improve the quality of content on the intranet.

Allocate some months (at least!) to the migration, as there is no shortcut to this process.

For more information on migration options and strategies, see the following article:

www.steptwo.com.au/papers/kmc_migration

Support content owners and authors

In a redesign project, the whole world can change for intranet authors and publishers. Establishing a new intranet for the first time can be equally disruptive for staff who are now expected to maintain content on the site.

From the perspective of an author, many changes can occur during these projects, including:

- new top-level site structure

- new approaches to structuring local content areas

- new page designs

- new publishing tools

- updated publishing standards

With the quality and currency of content primarily resting on the work of decentralised authors, they need to be supported through the transition period. (The intranet will not be a success if no-one publishes content to it.)

This support can take many forms:

- involvement throughout the design process

- early briefings on the new structure and designs

- training in any new intranet publishing tools

- engagement in the creation of new publishing standards and governance

- assistance with structuring local site areas

- hands-on support and advice during content migration

- training and mentoring in 'writing for the web'

- help-desk support from the intranet team during the intensive phases of the project

One of the most successful approaches can be to establish an 'authoring community' that brings together decentralised authors and content owners with the central team. Meeting regularly (ideally face-to-face), this community can establish good communications and mutual understanding.

Create the community early in the project, before any work has been done. This gives the group sufficient time to 'gel' before the hard work of the content migration hits everyone, as well as providing the intranet team with valuable support throughout the intranet project.

Figure 15-1: Highly professional marketing and educational materials can be produced to support the intranet launch. Materials courtesy of Aon.

Launch the new intranet

When it comes to launching (or relaunching) the new intranet, it doesn't pay to be shy! While some corporate cultures may mean a lower-key launch is appropriate, there is always a requirement to communicate details of the new site to staff. These communications should start at the outset of the project, and provide updates as progress is made. This culminates in the 'go-live' day, when the new intranet is switched on.

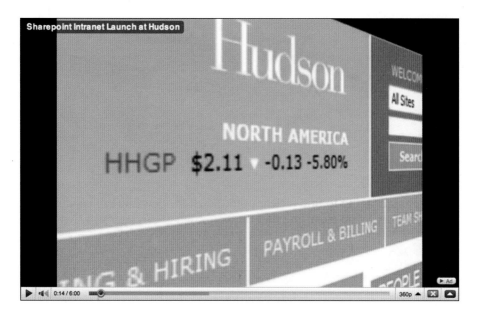

Figure 15-2: Videos can combine marketing and a guided tour of the new intranet. For the original video: http://www.youtube.com/watch?v=39052avrBlU

There are almost endless possibilities for an intranet launch campaign, including:

- intranet launch party or event

- intranet news items

- updates in internal newsletters or newspapers

- CEO message

- presentations to staff

- brochures (such as Figure 15-1)

- promotional items such as branded mousemats, stress balls, etc

- displays in the foyer

- posters in common areas

Intranet teams are also increasingly using video to promote the new intranet, combining marketing with a guided tour, such as Figure 15-2. Don't be afraid to be a bit playful when launching the intranet, although always match the prevailing culture of the organisation.

When to launch?

Content migration is no small task. For an intranet consisting of millions of pages (and these do exist!), it looks Herculean in scope. Carefully assessing and cleaning up each page before migrating it to the new site could take six months or more, even with the active participation of the decentralised content owners.

Time pressures and senior management expectations will often require intranet teams to draw a line in the sand, and to launch the new site before all content has been migrated or improved.

It can be challenging to find a clear point when the new site can be launched, and there may be no clear line that can be drawn around content that 'must' be migrated. (Where there are a separate 'satellite' intranet sites used by smaller audiences, these can often be safely excluded in the short-term.)

Follow these guidelines when determining when to launch:

- *Top-level pages have been improved.* The homepage and top-level sections of the intranet have been improved, with enhanced page layouts, structure and design.

- *Most-used sections have been migrated.* Key site sections such as HR and IT should be migrated, along with any other sections which are heavily used (such as forms, policies and procedures).

- *Delivers tangible and visible value.* The new intranet must be a concrete and substantial improvement, from a staff perspective. This should guide where work is done on the content, in terms of both cleaning up and migrating pages.

- *More than just behind-the-scenes.* Following on from the previous point, it must be remembered that staff and stakeholders don't see behind-the-scenes improvements such as a new publishing tool or metadata. It is therefore not sufficient to relaunch an intranet that looks the same, but works better for publishers and content owners.

- *Scope can be clearly communicated.* It must be possible to clearly communicate to staff 'we have improved x and y, but z will be done after launch'. This involves completing whole sections, and carefully thinking through how the remaining work can be described in simple terms.

There is only a single opportunity to officially launch or relaunch the intranet, so make the most of it. Even when a 'soft launch' approach is taken, there must enough delivered to demonstrate value and build enthusiasm. Use this yardstick to determine when to finalise the project and launch the intranet.

Help staff adapt to change

It is with great enthusiasm that the intranet team launches or relaunches the new intranet. The project has been hard, and the journey long. So there is an overwhelming sense of relief for the intranet team.

Unfortunately this may not be reflected, initially at least, in the reaction of staff to the new site. Staff often fall into two groups:

- *Loyal users of the old intranet.* These staff have learnt how to find the six things they use on the site, and are confused and disoriented when the new site is launched. Sad to say, the more staff use the intranet, the more unhappy they are likely to be when it changes.

- *Non-users of the old intranet.* These staff never picked up the habit of using the intranet, or abandoned it due to the difficulties in finding anything. Many of these staff won't even notice the relaunch, and as they don't use the intranet, won't appreciate the improvements.

Note that there is a negative outcome for both groups, and there may be few staff who are ecstatic about the changes. Content owners and key stakeholders also face disruption when the new site is launched.

Now this is the worst case, and many intranet projects will be flooded with praise and plaudits. It is worth, however, preparing for the worst and then being pleasantly surprised (rather than the other way around).

While good communication throughout the project can do much to mitigate resistance to change and launch problems, it cannot eliminate them entirely. Where concerns are legitimate, address them promptly and visibly. Where staff reactions are more emotional, weather them politely.

One organisation took an interesting approach to their intranet relaunch. When the new site was released, including a completely reworked homepage, a link was provided to the old homepage.

This was promoted as an option for staff who "didn't like" the new homepage, and a fair number switched back to the old design. Over the next few months, most of these 'holdouts' decided that the new design was actually better, and chose to return to the new design.

After three months, the old design was turned off. There were still a few resistant staff at this point, but they weren't going to be convinced regardless of the quality of the new intranet.

Giving staff the initial choice did much to reduce resistance to change, while still inevitably moving to the new design. These types of lateral approaches to intranet relaunches are always worth exploring.

Plan for the rest of the journey

While the intranet team deserves a well-earned rest after launching the new intranet, this is not the end of the intranet journey. There is no single *project* that delivers the 'perfect' intranet, and great sites are delivered via an ongoing *process*.

Every intranet project is limited by a range of constraints, such as budget, resources, time, support or technology. Carefully carried out, an intranet design or redesign project can make huge improvements. It will not, however, address every issue or deliver every required piece of functionality.

Staff usage after the launch will also shed light on issues (and mistakes) with the new design, so follow-on work will be required to gain the greatest value from the design project.

A process must also be put in place to sustain, maintain and grow the intranet. Too often, intranets are relaunched and then immediately start the slow slide back into disrepair, only to require a redesign three or four years later. (See the 'six phases of intranet evolution' in my other book, *What every intranet team should know* for more on this.)

Good governance will be required to ensure that content is owned and maintained. A permanent central intranet team (or individual) also needs to conduct ongoing improvements and enhancements. As staff usage and expectations evolve over time, so the intranet should change in sync.

The lifetime journey of intranets is beyond the scope of this book, and this will be covered in depth in a future volume. What can be done here is to highlight that the effort involved in designing or redesigning the intranet will come to little in the long term if not followed up with ongoing support and care.

Outcomes

The new site is live! Considerable improvements have been made to the intranet, and following a best-practice user-centred design methodology, the intranet team can be confident that the site works well for staff. Additional business benefits have also been delivered via the functionality that has been incorporated into the site alongside improved navigation and content.

Having completed the end-to-end design methodology, the rest of the book turns its attention to the many practical questions and issues surrounding intranet design and redesign projects.

Chapter 16

Accelerated design approaches

To this point, a robust methodology has been outlined for designing intranets. Involving staff in a structured way at each step of the process, intranet teams can be confident that the new site is effective and well-designed. If before-and-after testing has been done, it is also possible to demonstrate and quantify the improvements made.

This is not, however, a small piece of work. Following a best-practice methodology requires time, resources and budget. Skills and knowledge are also required, either within the intranet team, or through assistance from consultants or contractors.

The reality of many projects is that time, resource and budget constraints may make the full methodology impractical. It may also be excessive to conduct all the steps where the intranet or the organisation it serves is small.

Launching a new intranet where one hasn't existed before can also allow a more iterative approach to be taken that quickly launches a site and then immediately follows up with further improvements.

Intranet teams who find the full methodology daunting can follow a number of possible 'accelerated' approaches that cherry-pick key elements to deliver the greatest benefits within time and resource limits.

While spending more time designing the intranet will deliver better results, conducting a few user-centred activities is considerably better than none. Explore the accelerated options, and mix-and-match as required.

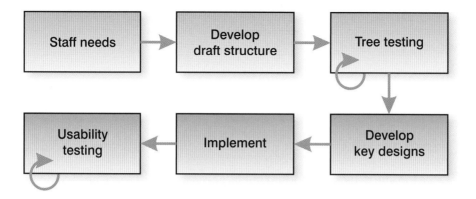

Figure 16-1: A targeted testing approach uses a small amount of tree testing and usability testing to refine the intranet before (and after) launch.

Option 1: Targeted testing

The targeted testing approach (Figure 16-1) takes the standard methodology, and prunes it down to use just two key techniques to test the design of the intranet. It works as follows:

1. Start by understanding *staff needs* (Chapter 4), if necessary informally. In just a day or two, it should be possible to get a sense of key needs, tasks and points of pain.

2. Develop a *draft structure* (Chapter 10) for the intranet. Draw upon all possible sources of information, including staff research, usage reports, previous surveys, best practices and real-world experience.

3. Use *tree testing* (Chapter 11) to quickly evaluate and refine the structure. Several days of work should be sufficient to identify and resolve major issues with the new site.

4. Sketch out on paper the *design for a few key pages* (Chapter 12). The layout of the content pages can be taken for granted, so focus on the homepage and perhaps a single landing page. Use rough paper sketches to refine the functionality that will be launched, and to check that labels make sense for staff.

5. Implement the site, and then conduct *usability testing* (Chapter 12) before launch. Conduct this testing as early as possible, so changes can be made quickly and easily.

This approach skips card sorting, relying on tree testing to match the new structure to how staff think and work. It shrinks down the page design phase, and uses usability testing later in the project to tune the implemented site.

Both tree testing and usability testing can be done in-house by the intranet team, even if there is limited experience with the techniques. (Getting outside professional assistance produce greater insight, if budget allows.)

The weakness of this accelerated approach is how late usability testing is left in the project. Since full draft designs are not produced, more comprehensive usability testing has to wait until a draft site is actually implemented. This may be too late to make important but larger changes, or the changes could be more costly.

Option 2: rapid prototyping

If a flexible and low-cost platform underpins the new intranet, a rapid proto-typing approach (Figure 16-2) can be taken. This quickly moves to an implemented intranet, at least in rough form, and then uses testing to quickly refine the designs. This works as follows:

1. Start by exploring *staff needs* (Chapter 4) as per the last approach, to determine where to focus the intranet, and to identify key staff tasks.

2. Develop a *fully or partially working prototype* that implements a draft structure and page layouts. Using a flexible intranet technology, this is quickly developed with a minimum of effort and cost.

3. Conduct repeated rounds of *usability testing* (Chapter 13) to identify issues, and to adjust the prototype. Re-test, and further refine, until an effective intranet is developed.

4. Migrate the rest of the content, and implement the remaining features leading up to the launch day.

This is potentially the quickest approach for developing or redeveloping an intranet. It does, however, rely completely on having an underlying technology platform that supports rapid prototyping and refinement.

This rules out many situations, where significant amounts of development or customisation are required to produce the intranet. Jumping straight into creating a draft intranet in these circumstances will produce a poorly thought-out design with little scope to change it.

(Many of the collaborative or social intranet tools are well-suited for this type of prototyping and experimentation.)

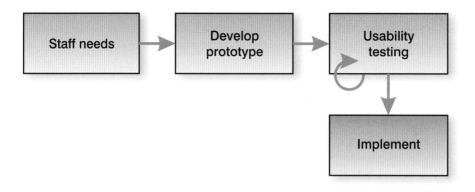

Figure 16-2: A flexible technology platform can allow a working prototype to be quickly created, and then refined iteratively via multiple rounds of usability testing.

Ensure you have excellent IT support if taking this approach, and be prepared to make major changes based on the results of the usability testing. This approach also puts significant pressure on the initial design implemented in the intranet tool, and on the effectiveness of the usability testing.

For this reason, it is valuable to obtain professional assistance with both steps, to ensure that the new intranet draws upon experience gained across many intranet projects.

Beware of this approach ending up as an IT-led project. Working directly with the technology doesn't eliminate the absolute requirement to meet staff needs, and to produce a site that is usable and intuitive. Instead, it increases the requirement for a strong user-centred approach, as there are fewer opportunities for testing with staff at earlier stages of the project.

Note that all of these approaches involving spending time initially to understand staff needs, desires and points of pain. Without at least some time spent on this, the new intranet can easily miss the mark, delivering functionality that is of interest to stakeholders (or the intranet team) but useless for staff.

Ensure that several days are spent on this research, at an absolute minimum. Considering the impact of the intranet on staff, a full week is strongly recommended.

Also take the opportunity to clarify the role and strategy for the intranet. The simplest way to do this is to determine the intranet brand (Chapter 5) at the outset of the project.

Option 3: quick-launching a new intranet

When creating a a brand-new intranet in an organisation, where one hasn't existed before, it can be hard to know where to start. Staff may have limited or no exposure to what an intranet even is, and by definition there isn't an existing site to evaluate and refine.

In this environment of uncertainty and opportunity, it can be useful to move as quickly as possible to launching an initial intranet. Deliberately designed to be small and simple, this site demonstrates the potential benefits of having an intranet, and provides a foundation for further expansion.

The first intranet that is launched can be considered a prototype for future enhancements, as well as being a working site in its own right. The key is to minimise the size of the site, while focusing on a few key features or content items that will be of benefit to the whole organisation.

If the site is simple enough, there may not even be a requirement for a full site structure, and this can be developed as the site grows. In this approach, usability and information architecture techniques are used *after* the launch, as content is added, rather than before.

(It definitely doesn't make sense to spend 12–18 months creating a brand-new intranet, if the organisation hasn't experienced the strengths and weaknesses of an existing site.)

It can be useful to thinking of this approach as a *funnel* (Figure 16-3). At the beginning are endless possibilities of what an intranet *could do*. The methodology then helps to narrow down the options:

1. Conducting *staff research* (Chapter 4) builds an understanding of current working practices and culture, and potential intranet opportunities.

2. The *intranet brand* (Chapter 5) is developed, providing a high level scope and purpose for the new intranet.

3. A more refined *scope* is produced listing functionality and content that will be launched.

4. *Draft designs and structures* are produced, which will be very simple for the initial intranet.

5. With scope and designs in place, the intranet is locked down, and the work moves to implementation and go-live.

Each step refines the focus of the intranet, producing concrete answers that inform the next step. Avoiding getting stuck in 'analysis paralysis', this approach quickly creates a site that grows rapidly after the initial launch.

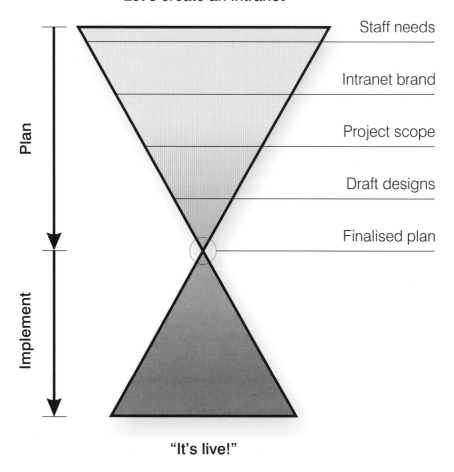

"Let's create an intranet"

Staff needs

Intranet brand

Project scope

Draft designs

Finalised plan

Plan

Implement

"It's live!"

Figure 16-3: Creating a brand-new intranet can be seen as a 'funnel', starting with a vague desire and narrowing down to a final design which can be implemented.

Beware of jumping in and creating a 'typical' intranet with core corporate content. This approach can also be excessively influenced by the desires or visions of a few key individuals, whose focus may not match the needs of staff.

For this approach to work well, the initial intranet must be concretely useful for staff, focusing on a few key features rather than a collection of documents or pages migrated from shared drives.

It must demonstrate what the intranet could be, and establish a framework that guides future additions. Even in this accelerated approach, careful planning and targeting of effort is critical.

Mix-and-match

Intranet teams around the world have independently created hundreds of variations on the core user-centred methodology for creating intranets. This includes:

- combined staff interviews and card sorting

- many different approaches to creating draft designs

- combined scoping and design activities

- collaborative approaches to the design process

- variations on core usability and information architecture techniques

When done well, these variations can perfectly fit the unique circumstances and environment of each intranet project. Don't be afraid to mix-and-match the techniques outlined in the previous chapters, or to use them in different ways.

The core goal is to deliver a site that works for staff. If this is always kept in mind, intranet teams are free to take many different approaches to the design process.

The purpose of this book is not to be proscriptive, or to burden teams with a methodology too extensive to be used in practice. Instead, take useful elements from the methodology, and apply them in ways that work in the real-life environment of your intranet project.

Outcomes

There are a variety of 'accelerated' approaches to designing an intranet, that use a smaller number of key techniques to create a workable and useful intranet. Apply these where time, resources or budget are limited.

Conducting some testing with staff is always better than none, and don't be afraid to experiment to find an approach that is the best fit for the project's needs.

Chapter 17

Personalisation and targeting

All staff needs are not the same. This is true in a 500-person local government agency, a 10,000-strong bank, or a global multinational with 100,000 staff. If the goal is to help staff do their jobs, intranets need to find ways of delivering content and tools that match specific staff needs.

This is inherently challenging (perhaps impossible) with a single corporate intranet. This leads many intranets to explore one, or both, of the following approaches:

- *User-driven personalisation,* where staff are given control over how the intranet works, allowing them to adapt functionality to meet their specific needs.

- *Targeting and tailoring,* where the intranet content and functionality is adapted or tailored on behalf of staff, based on what the organisation knows about staff.

There is considerable interest in these two approaches, driven in part by the rise of 'web 2.0' and 'enterprise 2.0' ideas. It is therefore worth looking at how they work, and what design options teams have.

The terminology in this area is a complete mess within the industry, with every vendor, consultant and expert calling these different things. All agree, however, on drawing a line between user-driven functionality, features driven by the organisation. Be aware of the confusion in terminology when discussing these features.

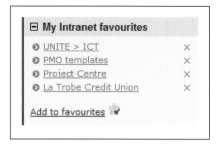

Figure 17-1: 'My links' or 'My favourites' is common functionality seen on many intranets. Screenshots courtesy of La Trobe University and Open University (UK).

User-driven personalisation

User-driven personalisation puts control of the intranet (to some degree) in the hands of staff. This feels instinctively to be the correct thing to do: we treat staff like adults rather than kids, and give them the power to adapt the intranet to fit their needs and working practice.

There are many variations on this concept. At the simplest level, it involves providing a 'my links' or 'my documents' feature, as shown in Figure 17-1. This is potentially helpful for staff, although its greatest value comes when staff can't use the browser's built-in bookmark feature (such as staff working in 'thin client' environments).

Beyond this, some intranets allow core features of the intranet to be personalised by staff. In Figure 17-2, for example, staff can subscribe to individual news feeds in addition to the main corporate news. Most features on homepages are open to this type of personalisation.

At the most extensive end, some intranets are established with fully personalised 'portal' homepages, such as Figures 17-3 and 17-4. These allow every aspect of the entry page to be configured, and staff can add their own 'widgets' or 'portlets' to the mix. In addition to providing news and navigation, these portal-style homepages often emphasise integration with back-end systems and other online tools.

Figure 17-2: In the highlighted area on this page, staff can personalise the news feeds they want to follow. Screenshot courtesy of COWI.

The technology to deliver this type of personalisation has become widely available, often out-of-the-box, at least for basic features. Web content management systems, portals and collaborative platforms all offer some degree of personalisation.

Personalisation can be very alluring to intranet teams. If staff can configure their own user interface, perhaps the methodology outlined in this book can be abandoned. Instead of the central team carefully crafting and testing what appears on the homepage, and how the intranet is structured underneath, it can be left to individual staff to meet their own needs.

When combined with features such as social bookmarking (page 126), perhaps this heralds a new way of delivering intranets? It is certainly a lot more 'sexy' than traditional intranet designs.

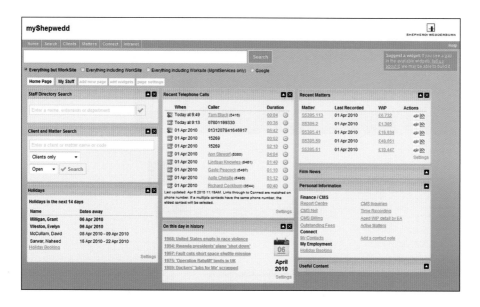

*Figure 17-3: A fully personalised page providing staff with key information and tools.
Screenshot courtesy of Shepherd and Wedderburn.*

When considering user-driven personalisation, however, there is one huge stumbling block that must be overcome in today's organisations:

Only 5–10 per cent of staff will make use of personalisation features.

Some readers may find this statement shocking. This finding has been observed across the globe, in all types of organisations, for a wide range of technologies. It has been confirmed by almost every intranet team, consultant, contractor and intranet expert.

Even the best-case situations, such as IT firms or technology-focused consultancies, have not exceeded this level of use. This is despite the enthusiasm for personalisation seen in both product vendors and many intranet teams.

In light of this, a fresh perspective on user-driven personalisation is needed.

What is perhaps equally surprising is how few intranet teams actually measure the uptake of these features when they are implemented. The vast majority of intranet teams have absolutely no information on how many staff have used the personalisation features even once, let alone how frequently or extensively they are being used now.

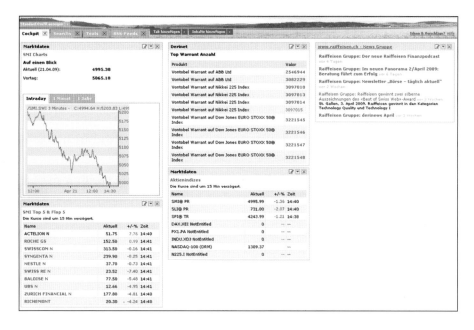

Figure 17-4: An intranet 'portal' interface developed by a Swiss Bank, and released as an open-source tool. Screenshot courtesy of Raiffeisen.

When confronted by the low staff take-up of personalisation features, intranet teams have a number of practical options:

1. *Avoid user-driven personalisation,* leaving it to a future version of the intranet when adoption and use is more certain. This allows time to be spent on other aspects of the site.

2. *Implement personalisation in a small way,* starting with a few simple pieces of functionality, and then expanding capabilities if there is sufficient staff interest and use.

3. *Ensure 5–10 per cent is considered a success,* by gaining the full value of the features even when only a small proportion of staff are using them. For example, real benefits may be gained if key groups such as sales staff or call centre staff make use of personalisation.

4. *Make personalisation vital to how staff work,* by incorporating it into business tasks and behaviours. There are a few examples of organisations which have done this, gaining near-universal use by making the functionality business-critical. (This is still uncharted territory for most projects.)

What doesn't make sense is to build a whole intranet strategy around person-alisation, or to make it the central focus of intranet design. In time, usage patterns may change, but the priority today must be to implement features that will be used by staff.

But what about …

When the somewhat startling 5–10 per cent figure is stated, there are normally a number of objections raised. Addressing the most common:

- *Generation Y will expect these features.* This is widely stated with respect to all sorts of 'enterprise 2.0' functionality, but it is extremely problematic. For a start, the concept of a single generation behaving in a consistent way is a myth, not supported by research results. Even if we wait long enough for sufficient Generation Y staff to enter an organisation, it is not clear that their individual practices will lead to widespread organisational change.

- *It works for iGoogle.* Many have used sites on the public web such as iGoogle, which work very well. The motivation for using these public sites is not the same as for intranets within organisations.

- *But I use it!* Many of the readers of this book, and the author himself, use personalisation features. We are not, however, representative of typical users. This is why there is a small level of adoption, but no steady increase after that.

- *Better usability will mean greater use.* Many earlier portal products had extremely poor usability, which was a significant barrier to use. Even well-designed personalisation features have, however, struggled to gain adoption.

- *Product <x> features it.* Some recently-released products make a big deal of personalisation, using it throughout the product. However, just because a vendor has included it doesn't mean that staff will use it in practice.

- *It's easy, so why not.* Simple features, such as 'my links' are often provided out-of-the-box by modern tools. While they are fun, they may add little business value and take up valuable screen real-estate.

None of this should discourage potential exploration or experimentation with personalisation features. Innovators, including those who provided screenshots for this chapter, are to be encouraged and supported. In time, these early adopters will undoubtedly identify successful approaches that can be adopted by the wider intranet community. In the meantime, most intranet teams should approach personalisation in a pragmatic way.

Targeting and tailoring

The other approach to meeting the specific needs of staff is via targeting and tailoring. In contrast to user-driven personalisation, this involves the intranet team modifying what is delivered *on behalf of* staff. Based on what is corporately known about staff, the intranet delivers different information or tools, with little or no active involvement from individual staff.

Figure 17-5 shows what this can look like. The highlighted regions are tailored based on the logged-in user, including their business unit and geographic location. In a global firm such as GE, this allows both global and local needs to be met (more on this in Chapter 19).

Tailoring can also be done on a smaller scale, such as modifying individual elements on the homepage based on login details. Figure 17-6, for example, shows local news tailored according to geographic location, alongside corporate news.

Tailoring need not be restricted to just the homepage of the intranet. In Figure 17-7, policies and procedures are modified according to the state where the staff member is located. This includes modifying the overall list of policies, as well as adapting specific wording within each policy based on regional differences.

In practice, there are three main ways of tailoring and targeting content, summarised in Table 17-1.

Tailoring according to geography tends to be the simplest option, as there are a limited number of regions, countries or offices. Geographic differences are also reasonably well-understood in most organisations, such as local office news versus company-wide news.

Targeting content to business units is the next easiest option. This can be done in many different ways, from a single box on the homepage delivering business-unit-specific information, to much broader tailoring of the whole site. Many organisations choose to limit this type of tailoring to a few key business units, such as the call centre or sales areas.

Targeting by job role is the most challenging approach. Across a whole organisation, there are likely to be hundreds of different job roles, with varying needs. Individual job roles may not be very well defined, with a single role such as 'Project officer' having many different responsibilities, or similar roles having different names.

Providing 'universal' tailoring or targeting based on job role is therefore unlikely to be practical. It is possible, however, to target specific groups of staff, and to deliver information and tools to meet their needs.

Figure 17-5: In a global organisation, targeted information allows a balance to be gained between global and local content. Screenshot courtesy of GE.

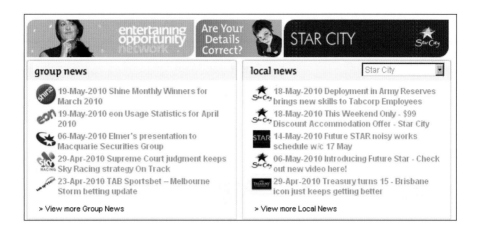

Figure 17-6: Local news is delivered alongside corporate news, tailored according to geographic location. Screenshot courtesy of Tabcorp.

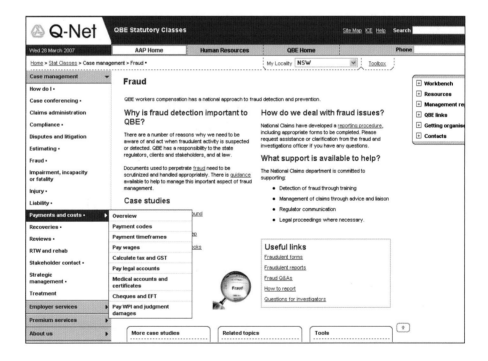

Figure 17-7: By choosing a 'locality' in the drop-down at the top of the page, insurance policies and procedures are tailored based on geographic location. Screenshot courtesy of QBE.

Target by	Difficulty	Description
Geography	Easy	Target based on the country, region or office that the staff member is located in. Can reflect differences in legislation or processes, as well as supporting the delivery of local news. This is easy for publishers and staff to understand, and geographic differences are comparatively simple to manage.
Business unit	Medium	Target according to the area of the organisation that a staff member belongs to. This can take place at many different levels, including division, business unit and team. This is relatively easy, as there are a limited number of business units, and their information needs are relatively well understood.
Staff role	Hard	Target based on the jobs that staff perform, recognising that different job roles have distinct information needs. This can be fairly easy if just a few roles are targeted (such as doctors and nurses), but is very difficult if attempting to cover an entire organisation due to a lack of clarity around formal job roles and descriptions.

Table 17-1: There are three main ways of targeting or tailoring information on intranets.

Targeting information to specific groups recognises that not all staff are equally important from an organisational standpoint. In any organisation, there will be specific groups who deliver most of the products or services. Examples include call centre staff, sales staff, production staff, and other operational groups.

Using the intranet to provide 'value add' for these groups can be relatively simple, while delivering substantial business benefits. This may be as small as providing additional links to key systems for frontline staff, or providing more extensive 'portal' views.

The good news about targeting and tailoring information is that it works. Unlike user-driven personalisation, the functionality is not dependent on individual staff choosing to modify their view of the intranet. By doing the work on behalf of staff, the intranet team can make the intranet meet local needs, as well as global ones. The simpler variants, such as targeting based on geography or business unit, are also usually simple from a technology stand-point.

There is, however, work to be done at the outset by the intranet team to establish targeting or tailoring. This is in addition to the time needed to complete the core design methodology outlined in the earlier chapters.

To be successful, intranet teams need to identify:

- key staff audience groups

- type of tailoring or targeting that will be delivered

- specific information needs of the targeted groups or locations

- page layouts and other design elements

- how it will be implemented in the underlying technology platform

- how the tailored content will be created and maintained

In general, tailoring and targeting content requires a greater level of intranet 'maturity', and it should be tackled once a strong intranet foundation has been put in place.

Address technology issues

A quick word about the technology required to deliver tailoring and targeting on the intranet. In order to deliver information based on geography, business unit or job role, the intranet needs to know key details about staff.

These details are typically drawn from underlying 'authentication' platforms run by IT. In simple terms, these contain the usernames and passwords staff use when they log on to their PCs each morning.

Over time, these details have been migrated into one of two standards: LDAP (an open industry standard) or Active Directory (the Microsoft variant of the same thing). These expanded 'directory services' have the ability to store much more than just names and passwords. They can also contain all the information that is in the internal phone directory or staff directory, including job titles, business unit names, locations and more.

LDAP or Active Directory is generally the source of the information needed to drive targeting and tailoring. Instead of the intranet building its own database of staff information, staff details are drawn out of this directory service platform.

Unfortunately, the information is not always stored in a useful way in these behind-the-scenes systems, which were originally set up for more narrow IT and security reasons. Even when information such as role and location is included in the system, the details may not be accurate or up to date.

While often very technical, underlying directory services are vital for the next generation of intranet and portal projects. The starting point for intranet and portal teams must therefore be to engage in discussions with IT about these platforms, to gain a clear understanding of what is in place, and how it works.

Any necessary improvements must start well in advance of the intranet work, and managed as a joint IT/business project. Only then will tailoring and targeting become possible.

Outcomes

Intranets are most valuable when they support day-to-day work of staff, but these activities vary greatly across job roles, business units and locations. Meeting all these needs with a single, static intranet is hard, perhaps impossible.

User-driven personalisation (where staff configure the intranet themselves) and targeting and tailoring (where information is delivered based on what is known about staff needs) provide a useful toolkit for intranet designers.

If a focus is maintained on what will be used and useful for staff, these approaches can help to deliver an even more successful intranet, going beyond 'flat' designs to provide a richer and more interactive experience.

Chapter 18

Design intranet functionality and applications

To this point, the focus has been on developing the top-down navigation of the intranet and overall page layouts. These activities have been about ensuring that staff can find the information they need to do their jobs.

The intranet is, however, more than just a collection of pages and documents. Core functions, such as search and the staff directory, are almost always the most-used aspects of intranets. These need to be well designed if they are to meet staff needs.

Intranets also provide gateways to a diverse range of business and frontline applications, supporting back-office and operational staff in their work. Increasingly, these applications are being incorporated seamlessly into the intranet, replacing links to free-standing web applications.

These applications, in their many different shapes and sizes, have the potential to deliver the clearest business benefits. Saving staff time is often a goal for intranets, and while this is hard to quantify for the content-rich areas of the sites, it is much more concrete for business systems.

Like every other aspect of intranets, user-centred design methodologies should be applied when creating applications. This chapter will touch upon the approaches that can be used, and the benefits of doing so.

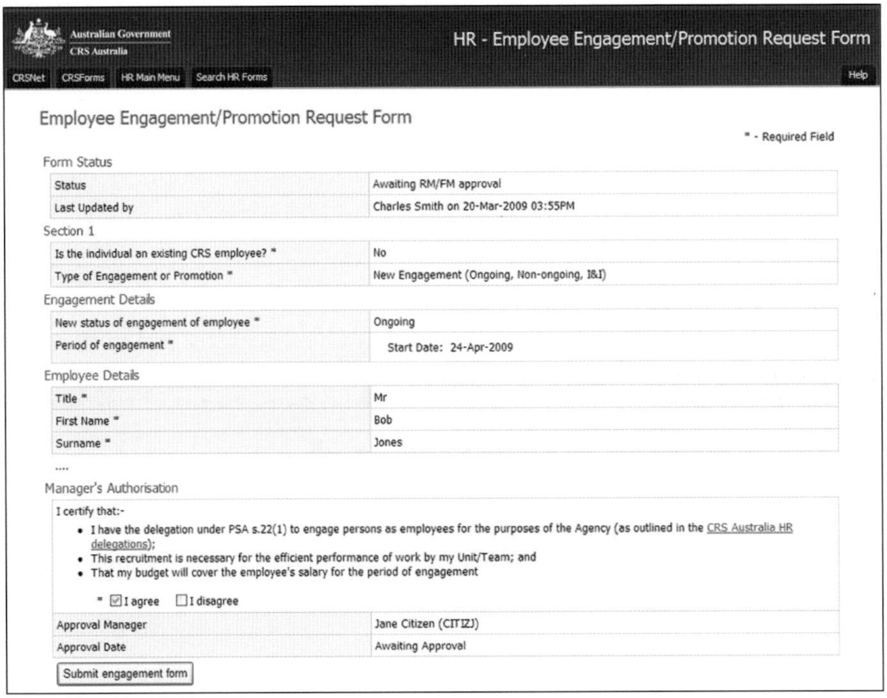

Australian Government
CRS Australia

HR - Employee Engagement/Promotion Request Form

CRSNet | CRSForms | HR Main Menu | Search HR Forms | Help

Employee Engagement/Promotion Request Form

* - Required Field

Form Status

Status	Awaiting RM/FM approval
Last Updated by	Charles Smith on 20-Mar-2009 03:55PM

Section 1

Is the individual an existing CRS employee? *	No
Type of Engagement or Promotion *	New Engagement (Ongoing, Non-ongoing, I&I)

Engagement Details

New status of engagement of employee *	Ongoing
Period of engagement *	Start Date: 24-Apr-2009

Employee Details

Title *	Mr
First Name *	Bob
Surname *	Jones

....

Manager's Authorisation

I certify that:-

- I have the delegation under PSA s.22(1) to engage persons as employees for the purposes of the Agency (as outlined in the CRS Australia HR delegations);
- This recruitment is necessary for the efficient performance of work by my Unit/Team; and
- That my budget will cover the employee's salary for the period of engagement

* ☑ I agree ☐ I disagree

Approval Manager	Jane Citizen (CITIZJ)
Approval Date	Awaiting Approval

Submit engagement form

Figure 18-1: Online forms that can be completed electronically can save considerable staff time and streamline business processes. Screenshot courtesy of CRS Australia.

Business applications

Intranets can be a 'place for doing things', in addition to a 'place for reading things'. While it is handy to have the leave policy in a location that can be easily found, it is even more useful to be able to apply for leave online.

Many intranets have the objective of being the 'gateway' to corporate information and tools, a 'single entry point' or a 'one-stop shop'. In the simplest form, this can be a list of links to internal applications, but there are many more opportunities beyond this basic goal.

Online forms are an obvious target on many intranets. While paper forms have been eliminated, the result is often a forms page containing a long list of PDF or Word forms. These have to be downloaded, filled in by hand, sent across by internal mail, typed in manually, and then processed.

These can often be replaced by true online forms, such as the one shown in Figure 18-1. These save staff time, and help to streamline business processes.

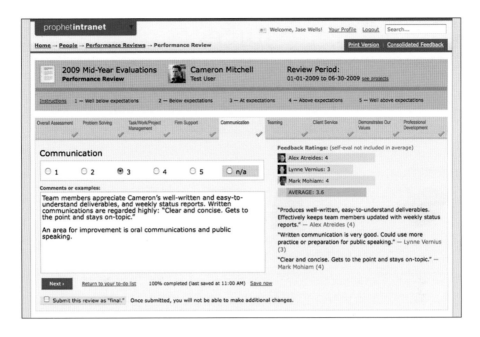

Figure 18-2: The firm's 360 degree evaluation process has been automated and streamlined in the intranet-based application. Screenshot courtesy of Prophet.

In practice, a 'rule of thirds' applies to intranet forms. Looking at the existing list of forms, the first third are simple forms that can be replaced with online equivalents using standard functionality provided by most content management systems and intranet platforms.

The middle third are more complex, requiring approval workflows or integration with back-end systems. These may require more powerful technology or some custom development. The final third are the domain of enterprise systems such as HR or finance systems, which are out of the scope for the intranet team beyond providing a simple link from the site.

Intranets can also be home to web-based applications, tightly integrated into the rest of the site. The possibilities here are endless, with most organisations having hundreds of processes and tools that would benefit from this type of development.

Figure 18-2 shows one example, where a 360 degree feedback process has been automated and streamlined in a consulting firm, reducing the turnaround time from months to weeks. Figure 18-3 shows an intranet application that allows staff to manage their own IT-provided equipment.

Figure 18-3: 'Juice' is an online application that allows staff to manage their company-provided equipment, including ordering upgrades and arranging replacements for broken items. Screenshot courtesy of Janssen-Cilag.

Design good applications

It is not enough just to deliver online versions of applications and processes. They must also work well for staff.

We have all, at one time or another, experienced frustration with poorly designed applications. Confused about where to click, baffled by the sequence of steps, we are prevented from completing our task by usability and design problems.

Unfortunately many internal applications suffer from these types of problems. Enterprise systems have long had a reputation for poor usability, creating staff frustration and generating inefficiency, rather than delivering the hoped-for productivity gains.

Intranets are increasingly expected to deliver tangible business benefits, such as streamlining key processes, saving money, or reducing transaction time.

These objectives will not be achieved simply by improving the quality of intranet content, and the ease of finding it.

The intranet as a 'business tool' must provide staff with functionality to support their day-to-day work. This goes beyond simply linking to existing applications from the intranet homepage.

This suggests a shift towards intranet-based functionality, and highlights the importance of good application design.

Good application design is therefore critical. In addition to providing the required functionality, the interface that is presented to staff must be easy to use and efficient.

This is where the field of *interaction design* comes in. This is a specialist aspect of usability that focuses on the design of applications, looking not just at how the pages are laid out, but also at the flow and sequence of user actions.

Drawing on experience gained from designing online tools and e-commerce sites, interaction design offers a practical toolkit that can greatly improve intranet-based solutions.

Techniques such as usability testing (Chapter 13) are core to producing applications that work for staff. There is also a growing set of design principles about how to present elements on the page, and common ways of completing tasks.

Ideally, the intranet team has a member with in-depth experience of interaction design, allowing the team to pursue with vigour the many opportunities for streamlining business processes. Alternatively, the team can draw upon skills elsewhere in the organisation, such as a specialist design team.

Where this isn't possible, it can be valuable to gain outside professional assistance when designing larger applications. This includes creating 'style guides' or 'design patterns' that can be used by both the intranet team and IT-based application developers.

Intranet search

To this point, the focus has been on delivering page layouts and site structure that ensure that staff can find what they need. As discussed on page 125, this is only half the story.

Search is a key component of a successful intranet. It is one of the most widely used intranet features, and is unfortunately also often the most disliked. With the huge breadth and depth of intranet content, search can either be a lifesaver for staff, or a hopeless time-waster.

There are many ways of obtaining search functionality: bundled as part of the CMS or intranet platform, purchased separately as a stand-alone search product, or built in-house using pre-existing components.

Search needs to work 'like magic'. Staff want to be able to type a few words into the search box, and to have the desired page or document appear in the first few hits. This doesn't happen without careful design, configuration and management.

The default search should be clean and simple for staff, with work done behind the scenes to make it function well. Where additional functionality is required, it should be provided in a straightforward and intuitive way. Figure 18-4 shows a clean design for intranet search that includes staff directory entries alongside content, and provides simple ways of narrowing down the long list of results.

Table 18-1 lists a few suggestions for improving the effectiveness of search. Most of these can be implemented using any modern search engine, with only a modest allocation of time and resources.

Search capabilities are constantly expanding, with a move towards 'enterprise search' that indexes other applications and systems. As the functionality and complexity of search grows, so does the imperative for good design and strong usability.

Improving Intranet Search
by James Robertson

It is beyond the scope of this book to do more than touch upon the importance and design of intranet search. For a full exploration, obtain a copy of Step Two's report on to this topic.

For more information:
www.steptwo.com.au/products/search

Figure 18-4: A modern implementation of intranet search, bringing together a mix of results into a single list, and providing staff with simple tools for narrowing down the results. Screenshots courtesy of AMP.

Eight ways of improving search

1. Tune to match intranet content and structure	Every intranet varies in terms of size, structure, mix of web pages and documents, and the use of metadata. Tune the out-of-the-box search engine to deliver useful and relevant results when searching current content.
2. Implement a simple default search	Provide staff with a simple search box and button, restricting any more complex functionality to specialist search users.
3. Simplify the results page	Strip out as much as 80% of the default functionality provided out-of-the-box by search vendors, to provide a clean, simple and usable design for general staff.
4. Track search usage	Establish two key search reports at a minimum: most popular searches, and failed searches (searches that return zero hits). Use this information to inform the design and structure of the intranet, and to further tune search.
5. Implement synonyms	Implement search engine synonyms so that terms such as 'bike' and 'bicycle' return the same results. This helps staff to find content even when they aren't using the 'correct' terminology.
6. Implement 'best bets'	'Best bets' ensure that key pages are presented at the top of the results, followed by the rest of the matches. For example, searching on 'leave' should return the leave policy and leave form at the top of the list.
7. Target the needs of specialist users	The default intranet search should be designed for 'typical' intranet users, searching for common content. Beyond this, specialist groups (such as researchers, lawyers, librarians, etc) will have more intensive needs. Deliver separate interfaces for these staff, rather than making the default search more complex for all.
8. Explore more advanced functionality	Once the core search solution is in place, explore more advanced functionality such as faceted search and enterprise search.

Table 18-1: A few of the practical steps that can be taken to improve the design and effectiveness of intranet search.

Staff directory

After search, the most used feature of many intranets is the online phone directory or staff directory. Providing a quick and easy way of finding phone numbers (and much more), this 'killer app' is much loved by staff. For this reason, the staff directory is typically given a prominent place on the homepage, as well as in the global navigation throughout the site.

Intranet staff directories provide core staff information such as:

- name

- phone numbers

- email address

- job title

- business unit

In most organisations, this is supplemented with further details such as:

- photo

- office location

- job responsibilities

- reporting structure (organisational structure)

- cost codes

Figure 18-5 shows a typical design for a staff directory entry containing a rich set of information. These additional details support a wide range of staff activities, beyond just finding a phone number or email address.

Increasingly, staff directories are evolving into business tools that integrate with other key systems. This allows details such as staff leave or workload to be displayed, as well as current projects and key deliverables. Staff directory information also underpins personalisation, content targeting, workflow and application functionality.

Staff directories are the natural hub for 'social' functionality within organisations. Staff profiles can show 'status updates', social and work connections, blog entries, and activities in collaborative spaces. With a focus on person-to-person connections, having a single staff directory to act as a glue between systems is critical.

Figure 18-6 shows an example of a staff directory page that incorporates both business functionality and social elements.

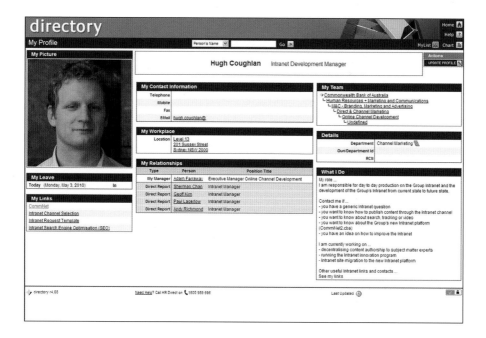

Figure 18-5: Typical design for a staff directory entry containing core contact details plus supporting information. Screenshot courtesy of Commonwealth Bank.

Staff directories add the greatest value when there is extensive cross-linking, both within the directory and to the rest of the intranet. Tightly integrating the directory as a single source of truth eliminates many opportunities for out-of-date information, as well as helping staff quickly find what they need regardless of where they are on the intranet.

For all these reasons, improving the staff directory is a common goal for many intranet design and redesign projects. To achieve this, a careful assessment must be made at the outset of the project of the technical and business process work required.

To be fully effective, staff directories must draw information from a number of systems and sources, including the HR application, core IT authentication systems, and staff-entered details. This often requires integration or application development, adding complexity and risk to the intranet project.

In most modern environments, staff directories are built on top of a platform such as Active Directory or LDAP. These provide the glue that connects systems, but are the domain of IT infrastructure areas. Approach these teams early to understand options and opportunities for enhancing the staff directory.

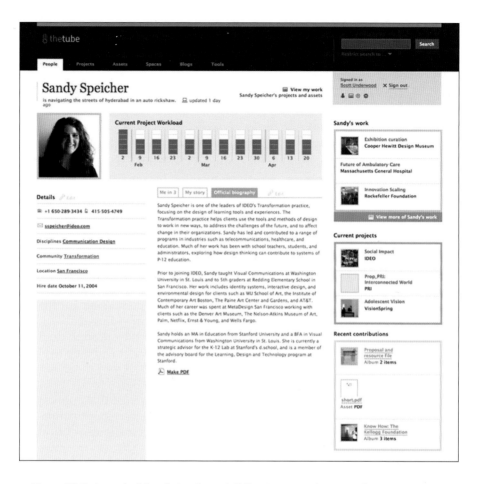

Figure 18-6: A much richer design for a staff directory page, incorporating core business information such as project workloads as well as social aspects. Screenshot courtesy of IDEO.

Staff Directories
by James Robertson and Donna Maurer

Intranet staff directories can be much more than just a collection of names and phone numbers as the past few pages have shown. This report provides hundreds of screenshots and insights on how to deliver a best-practice staff directory.

For more information:
www.steptwo.com.au/products/staffdirectories

Explore lines of responsibility

As discussed on page 157, there is a blurry line between the intranet and other internal applications and systems. This is also reflected in questions about ownership and responsibility for delivering intranet-based applications and processes.

Where the intranet is owned and managed by IT, responsibility for web-based applications will clearly sit within that group. Where the intranet is owned by communications or a business area, there may not be a mandate for application development within that group.

In either case, establish clear governance relating to intranet-based applications. Major systems, such as HR and finance, will always be the domain of IT. Smaller solutions may be produced by a developer within the intranet team, or in collaboration with IT.

Guidelines should be developed that outline the desired end-state for the intranet. These should emphasise the ultimate goal of providing staff with a seamless environment for all the tools and information they need to do their jobs. While this must necessarily be a long-term objective, it will help to align shorter-term activities.

It may also be useful to develop a set of basic design templates that application developers can use when creating solutions. Ensuring close coordination between the intranet team and developers early in the process is also critical.

There is no one-size-fits-all approach to managing intranet-based application development, and each organisation must find its own path. Whatever approach is taken however, must always place the needs of staff at the centre of activities, to ensure that usable and useful solutions are delivered.

Outcomes

As intranets shift towards becoming a 'place for doing things', the design of both core functionality and intranet-based applications becomes critical. The intranet will only deliver hoped-for efficiency gains if staff find online processes to be quick and easy.

The fundamental user-centred design methodologies apply equally well to applications and content. By understanding staff needs and testing with end users, intranet teams can deliver functionality that works well, saves time, and delivers tangible business benefits.

Chapter 19

Other practical considerations

The goal of this book is to provide a practical and approachable methodology for creating an intranet that works well for staff. Taking a step-by-step approach, a range of principles and techniques have been explored.

By necessity, the depth of information provided has been limited by the desire to avoid publishing a phonebook-sized tome, and the recognition that few intranet teams have time to read at length.

This has left a number of practical considerations unaddressed, however, and the purpose of this chapter is to cover a few of these. Here again, issues will be covered at a high level, giving pointers where appropriate to further information.

Many of the topics that will be discussed relate to more complex intranet challenges, beyond designing a straightforward site for a modest number of staff. Large organisations, and large intranets, have their own issues to address. Where intranets span a great diversity of staff or geographic locations, further planning and design effort will be required.

Successful intranet design must also draw upon overall intranet strategy, and consider how the site will be managed into the future. Design projects do not exist in isolation, and must be well-managed as projects.

Beyond the topics covered in this chapter, I encourage you to seek out intranet teams who have completed design projects, and to learn from their successes and failures. Many small activities make up the project as a whole, and each needs to be successful if the overall project is to deliver its objectives.

Avoid big-bang redesigns?

Designing or redesigning an entire intranet is a large project. When content migration is included, it will typically take no less than 6–18 months. Quite a few projects have taken two to three years to complete (ouch!).

Major intranet design or redesign projects are exhausting for intranet teams. Even when an accelerated methodology is used (Chapter 16), the demands of content migration still make it a large project.

Where this can be avoided, do so. Intranet teams often feel that they 'must' or 'should' redesign everything, but it often isn't the case.

Taking an incremental approach, using the same techniques and methodologies, can deliver a better result over time, without the pain of a major project.

It may be that a 'big-bang' intranet redesign is unavoidable. The current intranet may be suffering too many problems, or the move to a new technology platform may trigger a complete site rework. If so, the methodology outlined in this book will provide a solid foundation for the design project.

The fundamental difficulty with a complete redesign is that it aims to resolve all intranet problems in one project. Every project, however, is limited by a range of constraints. These typically include time, budget, resources, technology and internal politics.

The result is that, while the project delivers improvements, not all issues or needs are addressed. This becomes a problem when the work is done as a one-off big-bang project, and the site goes into 'business as usual' mode afterwards.

Experience in many organisations has shown that major redesigns may deliver only 50–75 per cent of hoped-for benefits, and then intranets go into maintenance mode. Issues and needs pile up over time, forcing another major redesign in three or four years.

The antidote to these issues is to take an incremental approach to intranet improvements. The same underlying user-centred design methodology is followed, complete with the key techniques such as card sorting, tree testing and usability testing.

The difference is that what was previously managed as a single project, is instead done in a series of individual improvements. This is drawn from the observation that many intranets suffer from an accumulation of many small issues, rather than one big problem.

There are many ways of slicing-and-dicing an intranet improvement. The HR section could be redesigned to deliver better navigation and landing page content. The 'useful tools' or 'policies and procedures' section could be made more useful.

High level changes can be made at the top of the site, including the homepage, and then propagated downwards in later projects. New features can be added to the site over time, and adjusted based on usage levels and patterns.

Taking an incremental approach limits the budget and resources needed at any given point, which should allow at least some changes to be made by the current team without extra support. It also allows lessons to be learnt from early changes, improving the design and delivery of later enhancements.

Even if a large project is required, follow these principles to incorporate an incremental approach:

- Wherever possible, break the overall intranet project into multiple pieces or stages, each of which is viable in its own right.

- Ensure that every project delivers tangible business and staff benefits, beyond addressing current site issues.

- Establish a culture of continuous improvement to the intranet, so that enhancements are steadily made over time, outside of major projects.

- Set the expectation with senior stakeholders and content owners that further improvements will be required after the initial go-live date.

- At a minimum, plan for a second round of improvements once the strengths and weaknesses of the new design have been fully experienced.

- Endeavour to allocate on-going budget to intranet improvements, beyond the up-front resources allocated to the major redesign project.

- Avoid unnecessary changes to the underlying technology platforms, as these can often trigger a major redesign, as well as creating a steep learning curve for the new technology.

Whatever the approach taken, ensure that it really will deliver the hoped-for benefits, and the site will continue to grow and evolve over time.

Quite a lot has now been written about the perils of 'big bang' redesigns, relating to both public-facing websites and staff-facing intranets. For further arguments and practical suggestions, read the following articles:

www.steptwo.com.au/papers/kmc_bigbang/

www.uie.com/articles/death_of_relaunch

www.uie.com/brainsparks/2006/08/10/more-on-why-major-relaunches-are-a-bad-idea/

www.gerrymcgovern.com/nt/2007/nt-2007-07-30-redesign.htm

Design for sustainability

Life doesn't stop when the new site is launched, and requests for additions and changes will continue to flood in. If the intranet is to avoid becoming messy and unusable again, thought needs to be put into long-term sustainability.

The new design and structure for the intranet can be developed with this specifically in mind. Consider the following guidelines and suggestions:

- *Design for expansion*, where it may be needed. For example, don't create a design that only supports six menu items across the global navigation if you're likely to add a seventh in the future. Ensure that the information architecture can accommodate new pages over time without breaking down.

- *Limit growth*, where appropriate. For example, create a small number of spaces on the homepage specifically for internal marketing, to prevent marketing messages from overwhelming the site over time.

- *Create clear distinctions* between intranet elements. Draw clear lines between the uses of different elements, so that when a new item or link needs to be added, everyone is clear where it should go. For example, having both 'quick links' and 'key applications' on the homepage will make it harder to manage the site over time.

- *Document the thinking* that went into the design. Creating a 'design justification' can help to capture the underlying principles that guided design decisions. Return to these when considering additions or changes to protect the integrity of the design over time.

- *Establish policies and governance.* Good design needs to be supported by clear governance and guidelines. These can be kept simple and informal if necessary, but can still help to avoid or mitigate common intranet challenges. The intranet homepage, for example, should always have a policy governing what is published to it, and what isn't.

- *Steadily re-test and refine* the intranet over time. When larger changes or additions are made to the site, take the opportunity to re-test relevant elements, using tree testing (Chapter 11) or usability testing (Chapter 13).

Fundamentally, there is no purpose in creating a 'perfect' design that works only for an instant, and is quickly overwhelmed and diluted by incremental change. Keep an eye to the future, and design the intranet to be flexible and adaptable where needed, and robust where required.

Consider global and local

Not all staff are the same. Located in different areas of the organisation, doing different jobs, their intranet needs will be quite distinct. The assumption has been, to this point, that a single intranet will be delivered that meets all staff needs. These greatly varying needs are not easy to meet.

The temptation is to focus on information that is relevant for all staff, with the goal of having the greatest impact. While this information has the greatest reach, it is often restricted to corporate services content such as HR, finance and IT. While this information is useful when needed, it is not directly related to the delivery of the organisation's services or products.

Information to specific business units or groups is more likely to be related to operational requirements, and tied directly into service delivery. This is the information that staff need every day to support their jobs. The challenge is that each staff group requires different information, which is hard for the intranet team to address within a single project.

To help with planning an intranet design project, it can be useful to divide information as follows:

* *Global (common)*: information or tools that are needed by the whole organisation, or the vast majority of staff.

* *Local (specific):* information or tools needed by a specific group of staff, according to location, business unit or role.

Both of these needs are valid, and must be met. It is not sufficient to focus on global needs, and to ignore operational requirements. Conversely, a fragmented collection of intranet sites addressing local needs provides no mechanism to communicate global news or content.

These challenges have already been touched upon throughout the user-centred design methodology. When developing a new site structure (Chapter 10), content was divided into three categories: core content (global), shop window and back office content (local). Personalisation and targeting (Chapter 17) provide mechanisms for meeting global and local needs, although they are not sufficient in isolation.

In the next section, the challenges of large organisations or very large intranets will be explored, and these have the global/local issue at their heart.

Intranet teams must make a commitment to meet local staff needs as part of the intranet redesign project if the site is to be used and useful. Use this principle throughout the project, from initial staff and stakeholder discussions, through to final page design and implementation.

Large organisations or large intranets

One large bank talks about their intranet 'solar system', with planets of content, themselves orbited by moons. Asteroid belts of rubble thread their way between the planets. Another large organisation has an 'intranet ecosystem' with a jungle of tall trees, with undergrowth and clearings.

What is clear is that intranets can get very large and very complex, particularly with large organisations. In these situations, it doesn't make sense to talk of a single 'intranet', rather a collection of intranet 'sites', connected together at the top level.

Like many aspects of intranets, these complex environments evolve over time. In the absence of an overall strategy or direction, new sites, sections, tools and applications were added to meet specific business needs. The purpose of these sites changed over time, some merged, others were deleted.

Tackling this environment with the user-centred design methodology outlined in Chapter 7 quickly runs aground. The global-local challenge quickly rears its head, and the complexities overwhelm the mental capacity of all involved.

What is needed in these cases is an overall 'intranet framework'. This defines the high-level shape and structure of the intranet landscape, and can answer questions such as:

- How many intranets will we have?

- What is the role and purpose of each element of the intranet(s)?

- What principles guide overall intranet decision making?

- What will be the homepage(s) for staff?

- How do we address global and local needs?

- How are communication messages propagated throughout the organisation?

- What role will technology play in the design of the intranet(s)?

These are all vital questions to answer at the outset of a project tackling a complex situation. Once the overall intranet framework has been determined, a user-centred design approach can be used to produce the page designs and information architecture.

In many cases, organisations have shied away from addressing these questions and determining an intranet framework in their redesign projects. The organic structure is left in place, and improvements made where possible.

In the absence of a clear framework, however, the benefits gained from the project are likely to be limited. Competing interests from stakeholders will make decision-making slow and uncertain. Legacy design decisions will impact on the usability (and utility) of the final site or sites.

In practice, there are many different possibilities for the top-level intranet framework. These include:

- Single intranet for the whole organisation, with personalisation and targeting to address local needs.

- Entirely separate sites for distinct business units, joined together by a global site that acts as a gateway.

- Global and common information brought together into a single task-based corporate intranet, supported by local intranets where specific needs have to be met (such as call centre and front-line environments).

- Single intranet with multiple branding, tailored based on the login of the staff member accessing the site.

- Syndication of global information into local sites.

Note that these options are not mutually exclusive, and many different permutations are possible.

Creating an intranet framework is first and foremost a stakeholder and business decision. Arising out of a consensus amongst key stakeholders, the framework is underpinned by agreement on the guiding principles for the intranet.

More detailed decisions can then be made by the intranet or project team, in consultation with an appropriate stakeholder group. Taking this top-down approach ensures that internal politics and differing priorities are addressed at the outset of the project.

The development of a intranet frameworks is still evolving practice in the intranet industry. Good work has been done by individual teams, but is not visible to the wider community. Best practices and guiding principles are still emerging.

Jane McConnell (*www.netjmc.com*) is a specialist in global intranets, and Step Two (*www.steptwo.com.au*) is actively working in this space. We will be publishing more on this in the future, and it will no doubt be expanded on in future editions of this book.

Multilingual and multicultural

To this point, this book has been written from the comfortable assumption that each organisation has a common language and shared national culture. When this is not the case, there are additional issues to take into consideration.

The core methodology is valid wherever it is applied, although it may need to be modified where there are multiple local cultures and different degrees of exposure and commitment to user-centred thinking.

Tools such as the Microsoft Production Reaction Cards (Chapter 5) can be easily translated as required. Staff research (Chapter 4) can be conducted in multiple languages, and the results consolidated into a single working language.

There are also multilingual and multicultural considerations when designing the intranet itself:

- *Languages used.* What languages will content be published in? What language will the navigation be in? Will the end user have a choice of languages? In a multinational organisation, this is determined by the overall business strategy and model.

  For example, common content may be published in the corporate 'working' language and specific content in local languages. The intranet home page itself, where global blends with local, may well have content side-by-side in two languages.

- *Cultural considerations.* Working practices, attitudes and behaviours vary greatly between national cultures when it comes to information management and collaboration styles. These differences will impact the information architecture, the style of the home page and the management of the collaborative spaces.

- *Project approaches.* Ways of engaging staff and stakeholders in the design project will vary according to cultures, and this must be carefully considered at the outset.

- *Intranet design.* The design and appearance of the intranet should match cultural norms and expectations. (For example, the colour red means very different things in China and the USA.)

To a certain extent, these issues are just another element of the challenges confronting teams working in large, complex organisations. In every project, care must be taken not to disenfranchise groups of staff, or take a 'head-office centric' view of the site. There is also no one 'right' way of designing an intranet, but rather the right way for the organisation's staff and culture.

Where do collaboration spaces fit?

Collaboration tools are spreading rapidly, driven by a new generation of intranet technologies, and the recognition that most day-to-day work involves staff working in teams and interacting with each other.

These tools take many forms, and are handled in different ways in the design process, as summarised in the following table:

Collaboration tool	Approach
Team spaces	The fastest growing type of collaboration tool, these provide local spaces for teams to share project information. To a large extent, these don't sit within the main site structure of the intranet. Access is provided to 'my team spaces' on the homepage (page 192)
Blogs	CEO and senior executive blogs may be showcased on the intranet homepage. Also consider including blog posts in news feeds or social updates on the intranet, as well as providing a specific section that brings together all the blogs into a single location.
Wikis	Wikis can be used as the basis for the whole intranet, as discussed in the coming section. When used as local collaboration spaces, they should be managed the same as team spaces, with similar approaches taken to access and site structure.
Discussion forums	These may be used by the whole organisation (such as buy-and-swap areas), in which case they should be given a suitable location in the overall site structure, and potentially on the intranet homepage. More localised discussions should sit alongside team spaces and wikis.
Mailing lists	The intranet should provide a list of all non-secure mailing lists, and a simple mechanism for subscribing or unsubscribing.

Whichever approach is taken, it's important to draw clear lines between publishing spaces and collaboration tools, and support this with simple but robust governance. This will mitigate the danger that the intranet and collaboration spaces will end up competing, with neither winning.

Figure 19-1: Intranets can be published using a wiki or 'social intranet' tool, putting collaboration at the heart of the site. Screenshot courtesy of REA Group.

Designing social intranets

Beyond integrating collaboration tools into a traditional intranet, some organisations have moved to a completely collaborative intranet, as shown in Figure 19-1. This may be published using a wiki, or one of the 'social intranet' tools that are gaining some prominence.

These tools 'democratise' intranet publishing and use, typically allowing any staff member to publish content or make changes. These tools commonly also take an 'all collaboration, all the time' approach, where the fundamental structure of the intranet is built around team and business unit collaboration.

Collaborative intranets are often seen in smaller or mid-sized organisations, where the cost and resources required to establish a corporate intranet are not practical or desirable.

These tools provide a number of advantages, particularly in the way that they engage with staff, and allow needs and structures to emerge over time. They also allow a new intranet to be quickly created with minimal cost or development.

The biggest stumbling block, however, is that information doesn't organise itself. As discussed earlier, tagging and folksonomies (page 126) can supplement traditional structures but not replace them. Collaboration and social connections are important, but don't replace the need for content and communication.

In the early stages of collaborative intranets, the emphasis is on building interest, awareness, enthusiasm and adoption. Using approaches such as intranet 'barn raising', staff are encouraged to find uses for the new technology.

Once the initial adoption phase has passed, the challenges of maintenance and growth set in. Left to their own devices, these collaboration-based intranets tend to become unstructured and messy. This parallels the organic growth seen in traditional intranets.

The design principles and methodologies outlined in this book apply equally to collaborative intranets as they do to every other intranet. To manage collaborative intranets, it is useful to return to the three types of content identified on page 116, and to use these to guide decisions:

- *Core staff and organisational content*: as in every other intranet, this information needs to be centralised into a task-based structure. This should be directly supported by a central intranet team, who help the content owners to take a user-centric view of their information. This is the most tightly managed content on the site, closer to traditional publishing models than open-to-all wiki-based editing.

- *Shop window content:* each business unit should be encouraged to provide a space in their wiki or collaborative area that is designed for an external audience, and structured accordingly. (Not every group will need a shop window space.)

- *Back office content:* this is where collaborative intranets really shine. By providing super-simple mechanisms for creating and updating content, and by seamlessly integrating this with collaborative and social functionality, these needs are well met out of the box.

The overall principle for collaborative intranets, like any other site, is to provide graduated governance and management. Core information needs to be correct, and should be more tightly managed and carefully structured.

As the intranet moves progressively outwards to meet local needs, light governance can be established, outlining basic guiding principles (such as appropriate usage) but otherwise leaving areas to manage their own content.

The challenge for the whole intranet industry is to find the right balance between two extremes. Traditional intranets can swing too far towards 'command and control' models, while collaborative and social intranets may not impose any structure at all.

Intranet teams are encouraged to avoid 'religious debates' when it comes to these issues. Over time, elements from both extremes will be brought together into more balanced technology platforms. Intranet teams need not, however, wait for vendors to lead these moves.

There are many opportunities to incorporate a mix of functionality into intranets, and intranet teams should not be afraid to experiment in this space.

The rise of 'social intranets' demonstrates that organisations having a growing number of technology options when it comes to creating an intranet. Traditionally, intranets relied solely on the capabilities of web content management systems (CMS) or enterprise portals.

As intranets mature, they are moving beyond just being publishing platforms for policies and procedures, and a channel for corporate news. Business tools and collaboration are becoming a particular focus.

Technology has evolved to match. The industry is much more diverse, with tools available that emphasis social and collaborative elements, and platforms focused on business processes and application development.

Intranet teams should examine a wide range of technology options, but be warned: with choice comes complexity! Teams may find choosing an intranet platform much more difficult, particularly when broader IT needs are considered. If in doubt, obtain expert assistance when evaluating technology options.

Compliance and regulatory issues

There are many compliance and regulatory issues that may affect the intranet, and the design project. As the site grows in usage and importance, so does the impact of these issues.

Issues that need to be considered vary greatly in different countries and types of organisation. Some things to consider:

- *Accessibility.* As a general principle, the intranet should be usable by staff with a wide range of disabilities, including vision impairment (blindness, colour blindness, etc) and motor control issues. There is an international standard for this (*www.w3.org/WAI*), and this is mandatory in some countries, and for some organisations (such as government agencies).

- *Privacy and data protection.* A number of countries, most notably in Europe, have strict laws regarding privacy and the storage of personal information. These may hold even when the organisation is headquartered outside these countries, but does business within them.

- *Financial legislation.* Organisations that provide financial services firms often have strict rules that cover staff activities, focusing on the information provided to customers or potential customers. Larger organisations will also fall under the remit of rules such as Sarbanes-Oxley, which may impact intranet activities.

- *Appropriate usage.* Staff sign up to strict codes of conduct and behavioural rules when joining an organisation. Processes must be put in place to ensure that these are not breached on the intranet.

- *Recordkeeping.* Legislation may require that information considered to be 'records' must be kept for a long period of time, and managed in a suitable system. This will often include documents on the intranet, and intranet pages themselves. Recordkeeping requirements vary between countries, but are likely to become more important as the role of the intranet increases.

- *Metadata.* Some organisations, most notably government agencies, are required to use a standardised set of metadata fields on all sites. While this is more typically mandated for public-facing sites, it may also impact intranets.

Other considerations may also apply to the intranet, depending on the specific circumstances. Make contact with the relevant business units to seek advice and appropriate strategies.

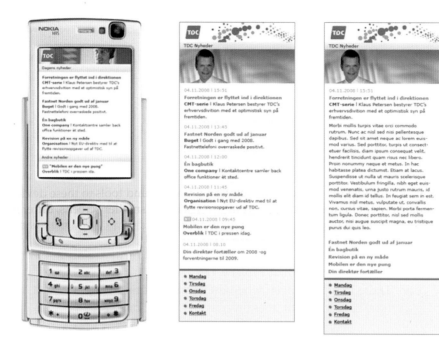

Figure 19-2: Intranet news delivered to mobile phones, providing a communications channel to staff in the field. Images courtesy of TDC Mobile.

Intranets beyond the desktop

To date, intranets have been used almost exclusively by staff sitting at desks, using a web browser on their PCs. Where field and operational staff have been considered, access may be granted via 'kiosks', although uptake and usage of these can be patchy.

There is no ignoring, however, the inexorable rise of mobile devices. With the new generation of devices offering rich 'apps' and web access, these are transforming how we gain access to information and tools.

The delivery of intranet functionality to these devices has been slow in coming. The diversity of devices used by staff, and the lack of a standard company-provided standard phone, have hampered uptake within organisations. Developing for mobile platforms has also been much harder than for desktop PCs. All of these difficulties are likely to ease, enabling easier delivery to mobile devices.

Early adopters have already started to deliver intranet functionality to mobile phones. Figure 19-2 shows intranet news on mobile 'smartphones', functionality that has been delivered by quite a few organisations.

Other organisations provide access to the staff directory on phones, a tremendously useful feature that will likely be taken for granted by staff in the near future. Content, forms and business systems are also useful on mobile devices.

Where organisations have a significant proportion of staff in frontline or operational environments, delivery to mobile device should be a priority. This includes staff in retail stores, on factory floors, or out in the field.

From an intranet design context, delivering content and tools to these devices brings new challenges and opportunities. There are smaller screens, different modes of interaction (no keyboards for example!) and new design metaphors.

In the short term, plan to experiment when designing for these devices, and draw on emerging best practices being developed for consumer applications. The use of third-party development platforms or delivery channels may also limit the impact of ever-changing mobile devices and capabilities.

Outcomes

An intranet has been launched that is easy and effective for staff. Careful planning has considered broader issues, including how staff needs will be met across the entire organisation. The sustainability of the intranet has been ensured by sound design decisions that provide the right mix of flexibility and robustness.

The result is an intranet that will grow and evolve over time, to match the constantly increasing expectations of staff, changing business needs, and the opportunities provided by new technologies.

Chapter 20

Further resources

Intranets are a big field, sitting at the intersection of many different disciplines, including information management, usability, communications, IT, change management, knowledge management and more.

This book has taken one slice through the field, focusing on the *design* of intranets. Teams are encouraged to read widely, to ensure success across all aspects of intranets.

Step Two Designs

www.steptwo.com.au

The consulting business that I lead has been working in the intranet space for over a decade. In addition to the reports and articles referenced throughout the book, there are literally hundreds of free articles for the taking. We also publish three new articles every month (except January), and you can sign up to be notified:

www.steptwo.com.au/subscribe

In addition to consulting, we run training sessions and workshops around the globe, as well as mentoring for intranet teams (face-to-face and remotely).

Intranet blogs and forums

- Alex Manchester (*www.alexmanchester.com*), one of the team at Step Two and a regular blogger.

- Column Two (*www.steptwo.com.au/columntwo*), written by me (!).

- Globally Local… Locally Global (*www.netjmc.com/blog*), written by Jane McConnell, the undisputed thought leader on global intranets.

- Intranet Blog (*www.intranetblog.com*), written by Toby Ward in Canada, one of the leaders in the global intranet community.

- Intranet Experience (*intranetexperience.com/ourblog*), providing practical advice on intranets from a number of contributors.

- Intranet Life (*www.intranetlife.com*), published by the Intranet Benchmarking Forum in the UK.

- Intranet Focus (*www.intranetfocus.com/blog*), written by Martin White, a UK-based intranet expert and author.

- Intranet Professionals (*www.linkedin.com/groups?gid=113656*), the most active LinkedIn group for intranet teams and other professionals.

- Mark Morrell (*markmorrell.wordpress.com*), regularly sharing insights from the intranet manager at BT (British Telecom).

- Worldwide Intranet Challenge (*cibasolutions.typepad.com*), sharing the results of an ongoing global benchmarking project looking at intranets.

Intranet communities

- Intranet Benchmarking Forum (*www.ibforum.com*), supports global intranet teams, particularly those in the Europe and North America.

- Intranet Leadership Forum (*www.steptwo.com.au/ilf*), based in Australia, with chapters in Brisbane, Canberra, Melbourne and Sydney.

- IntraTeam (*www.intrateam.dk*), supports intranet teams in Denmark and Sweden.

- J. Boye (*www.jboye.com*), supports a range of web professionals including intranet teams, in Denmark, the UK and the rest of Europe.

Usability and design resources

- Boxes and Arrows (*www.boxesandarrows.com*), a definitive collection of articles on information architecture.

- Gerry McGovern (*www.gerrymcgovern.com*), author of numerous books on web content, and leader in the intranet community.

- IA Institute (*www.iainstitute.org*), a membership organisation for information architects.

- Nielsen Norman Group (*www.nngroup.com*), led by Jakob Nielsen, one of the leading usability experts and publishers of a range of usability reports and the widely read intranet design annuals.

- Rosenfeld Media (*www.rosenfeldmedia.com*), run by one of the founders of information architecture, a publisher that specialises in books on user experience.

- Usability.gov (*www.usability.gov*), published by the US government, a comprehensive source of usability reference material.

- Usability Professionals Association (*www.upassoc.org*), a membership organisation for usability practitioners.

- User Interface Engineering (*www.uie.com*), led by Jared Spool, are leading usability experts and publishers of a range of reports and resources.

- UXmatters (*www.uxmatters.com*), an excellent online magazine publishing regular articles on a range of practical topics.

Index

W

Other Step Two publications

What every intranet team should know

This is the definitive 'quick start' guide to intranets, providing intranet teams with a to-the-point overview of how to plan, design, manage and grow intranets.

A beautifully printed A5-sized 110-page book, this volume covers key topics for every intranet team:

- Six phases of intranet evolution

- Four purposes of the intranet

- How to find out what staff need

- How to design the intranet

- How to deliver great content

- The role of the intranet team

- How to plan intranet improvements

Few teams have time to read a weighty tome on intranets, and what is needed is a clear 'map' for delivering a successful intranet. Drawing on experience from intranet teams across the globe, every page of this book provides key insights, ideas, models and methodologies.

To obtain a copy:
www.steptwo.com.au/products/everyteam

Intranet Innovation Awards

Since 2007, the Intranet Innovation Awards have celebrated new ideas and innovative approaches to the enhancement and delivery of intranets. These are the premier awards for intranet teams, and a valuable resource for the whole community.

Uniquely, these awards recognise individual intranet improvements, and not intranets as a whole. The awards are about improving all intranets, by sharing great ideas and increasing the pace of innovation across the whole of the intranet community.

Every idea, no matter how small, adds to our understanding of what it means to have a successful intranet.

With winners across four categories (core functionality, communication and collaboration, frontline delivery and business solutions), there are valuable ideas for every intranet team.

The awards are global in nature, with winners from Australia, Canada, Denmark, Italy, The Netherlands, Switzerland, Russia, New Zealand, UK and USA.

Each year, a report is produced that shares the full results of the awards, including screenshots and details of the winning entries. Commended entries provide further insights and examples.

Use these reports to discover where intranets are innovating, and to find ideas that can be implemented on your intranet.

Once you have used this book to deliver a great intranet, enter your site in the awards and win a beautiful glass trophy!

To obtain a copy of the reports:
www.steptwo.com.au/iia

Intranet Roadmap

The Intranet Roadmap outlines all the activities needed to design or redesign an intranet. It provides a practical project management methodology and checklist for teams approaching this sometimes daunting task.

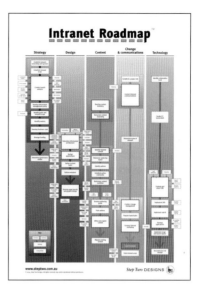

Our best-selling resource, the wallchart can be found on the walls of over 500 organisations, providing a starting point for intranet redevelopment projects.

Beyond just implementing software or redesigning the site, the Intranet Roadmap covers activities in five key streams:

- strategy

- design

- content

- change & communications

- technology

The Intranet Roadmap is delivered in two forms:

- full colour A1 wallchart

- supporting 54 page booklet

The wallchart lists the key activities required in each of the project streams. It also highlights which activities (such as usability testing, affinity diagramming, personas and collaborative design) can be used to support individual activities.

The supporting booklet then provides an overview of every activity and technique listed on the Intranet Roadmap, as well as linking to further resources and information.

The combination of the wallchart and booklet will be invaluable for any team looking to develop or redevelop an intranet, and it will assist in both planning and reviewing the approach taken.

To obtain a copy:
www.steptwo.com.au/products/roadmap

6x2 methodology for intranets

6x2 methodology for intranets

James Robertson
FEBRUARY 2007

Step Two DESIGNS
www.steptwo.com.au

Intranets must succeed. The organisational need to improve business processes, share knowledge and support staff in their daily work is constantly growing. Most organisations have already realised that they cannot operate without an intranet (whatever it might be called).

Yet the real challenge is for intranet teams to gain the support and resources to meet these business needs and demands. The net result is that intranet teams often work hard from month to month, but may make little apparent progress on the longer-term objectives for the site.

The 6x2 methodology works within these constraints to provide a practical approach for intranet planning that simultaneously delivers additional intranet functionality while building support for the intranet team.

At the heart of the 6x2 methodology is a focus on the coming six months. More than just steadily working on longer-term activities, this approach asks: what can be delivered in the next six months? In answering this question, it also identifies those activities that are not just doable, but also worth doing.

The possible activities are then sketched out for the following six months (thus the '6x2' name), giving the intranet team a roadmap for the coming year.

This is a cyclic process, with each six month period of activity leading into the next. Underlying this is the steady building of momentum for the intranet, giving an 'upwards spiral' that allows more to be done in each six months period.

This methodology provides a simple and pragmatic approach that can be used by intranet teams of any size (from one person to a dozen or more). It is equally applicable in private and public organisations, and the more complex and difficult the intranet, the better the approach works.

To obtain a copy:
www.steptwo.com.au/products/6x2

Improving intranet search

Poor search is one of the greatest sources of user frustration with intranets. Worse yet, the inadequacies of search may be consigning the intranet as a whole to failure.

The problem is that few search solutions have been effectively designed, in terms of the interfaces provided to staff or the behind-the-scenes improvements needed to make search work like magic.

Even just a few days devoted to improving search can have significant impact. If resources are then allocated to steadily maintaining and enhancing search, it will not take long to deliver an extraordinary search solution.

Improving Intranet Search

James Robertson
FEBRUARY 2006

Step Two DESIGNS
www.steptwo.com.au

In this 115-page report you will find:

- key principles of effective search

- nine step methodology for improving search

- two search 'personas' describing common search users

- guidelines for improving the search interface

- guidelines for enhancing search results

- options for search engine tuning

- introduction of search usage reports

- exploration of search engine synonyms and 'best bets'

- summary of advanced search techniques

- screenshots and tips throughout

Use this report to make immediate improvements to intranet search, and to then plan a program of work that will deliver a powerful search solution for staff.

To obtain a copy:
www.steptwo.com.au/products/search

Staff directories

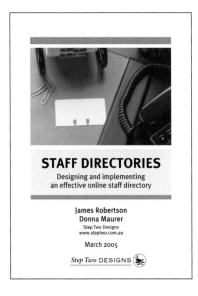

STAFF DIRECTORIES
Designing and implementing
an effective online staff directory

James Robertson
Donna Maurer
Step Two Designs
www.steptwo.com.au

March 2005

Step Two DESIGNS

Also known as phone directories, phone lists and corporate whitepages, staff directories are always the most-used feature of corporate intranets. More than any other tool, they are used every day throughout the organisation.

Staff directories can be much more than just a list of names and phone numbers. They can capture organisational structure, locations, photos, skills and expertise, projects, blogs and much more.

Like any tool, however, staff directories must be carefully designed to be effective and usable. This "better practice" report is designed to capture the experience gained across dozens of organisations to give you clear and practical ideas on how to design, implement and maintain your staff directory.

In this 91-page report you will find:

- detailed exploration of staff directory fields and features (both common and advanced)

- design guidelines for all staff directory pages (including search screens, profile pages, organisational charts, and more)

- in-depth discussion on how to keep your staff directory up-to-date

- outline of practical approaches to developing your staff directory

- full results of a public survey into staff directory usage

- tips and suggestions relating to all aspects of staff directories

- many screenshots and examples throughout

Anyone looking to develop a new staff directory, or improve the one they have, would benefit from this report.

To obtain a copy:
www.steptwo.com.au/products/staffdirectories